A History of Women in
Mathematics

For Professor Shirley Yap,
Who Showed me my Way
Back to Mathematics.

A History of Women in
Mathematics

Exploring the Trailblazers of STEM

Dale DeBakcsy

PEN & SWORD
HISTORY

AN IMPRINT OF PEN & SWORD BOOKS LTD.
YORKSHIRE – PHILADELPHIA

First published in Great Britain in 2023 by
PEN AND SWORD HISTORY
An imprint of
Pen & Sword Books Ltd
Yorkshire – Philadelphia

ISBN 978 1 39905 651 9

A CIP catalogue record for this book is available from the British Library.

Typeset in Times New Roman 11/13.5 by
SJmagic DESIGN SERVICES, India.
Printed and bound in the UK by CPI Group (UK) Ltd.

Pen & Sword Books Limited incorporates the imprints of Atlas, Archaeology,
Aviation, Discovery, Family History, Fiction, History, Maritime, Military,
Military Classics, Politics, After the Battle, Select, Transport, True Crime, Air
World, Frontline Publishing, Leo Cooper, Remember When, Seaforth Publishing,
The Praetorian Press, Wharncliffe Local History, Wharncliffe Transport,
Wharncliffe True Crime and White Owl.

For a complete list of Pen & Sword titles please contact
PEN & SWORD BOOKS LIMITED
George House, Units 12 & 13, Beevor Street, Off Pontefract Road,
Barnsley, South Yorkshire, S71 1HN, England
E-mail: enquiries@pen-and-sword.co.uk
Website: www.pen-and-sword.co.uk

or

PEN AND SWORD BOOKS
1950 Lawrence Rd, Havertown, PA 19083, USA
E-mail: uspen-and-sword@casematepublishers.com
Website: www.penandswordbooks.com

Contents

Contents

A Note on Inclusion

A Note on Inclusion, or 'Why Isn't [My Favourite Person] in this Book, DALE?'

So, you have scanned through the table of contents, maybe glanced through the Brief Portraits sections, and are flabbergasted, downright puzzled, as to why X is featured in a main article, Y is only in a Brief Portrait, and Z doesn't seem to be around at all. 'What is the deal, Dale? Clearly, Y should be a main, X should only be a Brief Portrait, and Z's absence is an affront to the entire history of mathematics. You fool.'

That last part is definitely fair, but as to Z's absence, there are one of two reasons for it, the most likely of which is that I am saving her for another volume. For example, Irmgard Flügge-Lotz and Hilda Lyon will be showing up in the Engineering volume, while Grace Hopper and Ada Lovelace will both be appearing in the Computer Science volume. The other potential reason, of course, is that I just missed her somehow, in which case, drop me a note on Twitter (if it still exists by the time this book is published) – she is probably great and it will be delightful to learn about her, and then she can make her appearance in the second edition, if and when I get to put that out. I love it when new recommendations for scientists to cover pop up in my mailbox or on my Twitter feed, so please poke me if your favourite person isn't here and doesn't have a second field of accomplishment under which category I might have stuck her.

As to which people are Xs and which are Ys, if I knew how my brain picked which figures to devote full articles to and which to summarise in brief portraits, I'd share that with you, but I don't, so I can't, so if you think I have committed a grave injustice in one of my choices, I ask your indulgence, and if you know of a source that I missed that might promote a Y to an X, again, just drop me a line!

Now, grab a lap cat and a mind-expanding energy drink, settle in, and get ready for a journey of infinite scope, brought to us by a cavalcade of minds profound in their genius, resolute in their determination, and unyielding in their rigour.

Introduction

For 2,000 years and more, women have been contributing substantively and steadily to the study of mathematics. In ancient Greece and Rome, at the heart of the Islamic Empire, through the depths of the French Revolution, and during the height of the British Industrial Age, women were attaining measures of renown in mathematics that were generally denied them in other fields, which leads us to the inevitable question: why has mathematics in particular been such a consistent field of achievement for women?

Many of the reasons for women's long-term representation in mathematics can be reduced to simple questions of optimising intellectual opportunity within the confines of a set of arbitrary restrictions. Mathematics does not require travel to exotic locations and discussion with local experts like early natural history did, and was therefore more accessible to a slice of humanity whose travel was strictly monitored and curtailed. In ancient Greece, women tended to marry around the age of 14, and once married, their travel was restricted to their circle of friends and attendance at religious festivals, and in ancient Rome, mobility for a married woman was hardly much improved.

Whereas travel was a necessary component of ancient biological studies, to witness for one's self variations in plant and animal life from region to region, and to speak with farmers and animal breeders about their work, mathematics could (and still can) be done more or less entirely from within the confines of a relatively quiet room. In addition, mathematical research can be carried out with a minimum of specialised apparatus. As physical science moved into the seventeenth century, research into physical and chemical phenomena was increasingly restricted to those who could afford expensive speciality instruments, meaning that the scientific class tended to comprise gentleman researchers who felt justified in investing in such devices for themselves but rarely did so for their spouses (though they would often admit those spouses into the laboratory as helpers, as we saw in the astronomy volume of this series).

In addition to the practical considerations of mathematics' accessibility for those without access to travel or specialised instrumentation, there were significant social aspects that nudged women of intellectual talent into mathematical studies in antiquity and beyond. Household management, which, for much of Western history, was women's most important role behind that of child production, required oversight of domestic inventories and budgets that necessitated at least some familiarity with mathematical fundamentals. Well into the modern age, basic mathematical knowledge has stood next to sewing, cooking, music and languages as the essence of a young woman's education, even as chemistry, physics and astronomy were viewed more often than not as either irrelevantly unnecessary or downright unseemly.

Beyond its utility in the domestic sphere, mathematics was also viewed, particularly in the nineteenth century, as the 'safe' science; as a branch of intellectual inquiry that would not lead to any dangerous, anti-Victorian, ideas. Biology, with its evolutionary thinking, was believed to lead inevitably to atheism and immorality. Medicine brought young women into contact with the brute facts of existence, which would damage the sublimity of their natures and irreversibly assault their sense of modesty and propriety. Anthropology required travel among potentially dangerous persons, and brought up questions of cultural relativism that might challenge the dominant racial paradigms. By contrast, mathematics was considered so abstract in its principles, and benign in its posed problems, that a father or husband could allow his daughter or wife to commit herself to its perplexities without worrying about her developing controversial opinions about humanity's place in the animal kingdom, or her race and class's position in the social hierarchy.

Was women's prominence in mathematics, then, arrived at entirely subtractively? Was it simply the only science left after you took into consideration what society deemed safe for you to know; what your sharply conscribed life responsibilities required you to know, what access to specialised equipment you had, what social circles you were allowed to circulate among, and what restrictions were placed on where you could move?

Well, no. As we'll see in future volumes of this series, there were other options open to women throughout history that remained within the confines of those social and practical restrictions. But more than that, the reason many of the women we shall meet in the coming pages chose mathematics was less out of a feeling that it was their only option, and more because of larger intellectual trends that made them actively move towards maths, rather than falling back into it. Theano of Croton was part of a Pythagorean sect whose interest in numbers and geometry was part and parcel of a larger view of the nature of reality, which carried with it attractive religious and social ideas. For Hypatia of Alexandria, mathematics represented an island of reason and tradition in the midst of a world devolving into religious zealotry. Émilie du Châtelet viewed mathematics as the keystone of the entire Enlightenment project, the tool employed by Isaac Newton to reveal the ultimate order of the universe and thereby place measurement and reason, rather than superstition or myth, at the centre of our conception of the cosmos. Pelageya Polubarinova-Kochina placed mathematics in the service of engineering as part of a larger life goal to make herself useful to her fellow citizens of the Soviet Union, while for Fan Chung the power of combinatorics represented a means of understanding the complex systems produced by capitalist cultures.

Mathematics has a wondrous omni-utility to it, then, which allows people of vastly different religious principles and civilisational values to find in it the tools they need to realise their particular goals. And perhaps it is the combination of this chameleon-like ability to represent all things to all people, with its capacity to work around even the most oppressive of social restrictions, that has made mathematics not only a fallback intellectual pursuit but a consuming life's work for so many women from such vastly diverse backgrounds, over the course of the last two and a half millennia. While not always accepted by the larger mathematical community (as we shall see in the struggles of Sophie Germain) or given full scope to achieve what they might due to unsupportive partners (as was the case with Grace Chisholm Young and the first husband of Mary Somerville), it is, nonetheless, a remarkable circumstance unseen in virtually every other field of science that, for every era in history, if mathematics has been a major pursuit of a culture, there have been women amongst the first rank of its practitioners. Let us meet some of them now.

Chapter 1

Theano of Croton and the Pythagorean Women of Ancient Greece

In a small but soon-to-be-revered town in southern Italy, 2,500 years ago, a group of men and women gathered, united by the proposition that the universe is, at its base, Numbers. They were called the Pythagoreans, and their society would last for a millennium while their mathematical discoveries will be part of every geometry textbook in every school for as long as there are humans to read them. At the centre of that tight-knit society were two people: Pythagoras himself, and a woman named Theano.

Theano was one of seventeen early women Pythagoreans mentioned by name in the historical record, and the only one about whom we have anything approaching definite to say, for between the foundational Pythagoreans and us lie a number of bedevilling filters and documentational chasms. Firstly, the cult of silence that formed a central part of Pythagorean practice meant that few of the early practitioners set their thoughts to paper. Secondly, the appropriation of Pythagoreanism by later Platonic philosophers who cavalierly recast its origin in their own image means that we have to tread carefully in separating Pythagorean thought as it has come down to us from its various Platonic encrustments. Thirdly, the fifth-century BCE breakup of the Pythagoreans into two rival camps (the Aphorists and Scientists) and their mutual intellectual decimation makes it difficult to determine the original belief system of that first generation of thinkers. And lastly, the catastrophic loss of ancient sources during the Christian Era has reduced the original texts of those rare early Pythagoreans, who *did* write down their thoughts, to lists of titles recalled in sources from centuries later.

With so many conspiring sources muddying the historical waters, it is a wonder we know anything at all of Pythagoras and his generation, but some basics seem assured. That women were welcome to practise Pythagorean philosophy, and to attain some renown thereby, is clear from the number of sources listing their names and works. The problem comes when we start

1

speculating on the contents of their work for, to a Pythagorean, 'philosophy' meant a good deal more than it does to modern ears. An ancient philosopher pondered questions of ethics, metaphysics, natural science, rhetoric and mathematics as part of their calling, and a Pythagorean philosopher in particular was a mixture of mathematician, theologist and ethicist, which is difficult to separate into its component parts.

When Pythagoras journeyed through the Egyptian and Babylonian Empires seeking wisdom, he came in contact with principles of reincarnation, lifestyle and the sacredness of Number, which he brought back to Greece and fashioned into an influential new philosophy that preached silence, lack of ostentation, a simple diet, the migration of the soul between bodies, the eternal recurrence of all things, and the basic numerical nature of the world. In devotion to this last point, the early Pythagoreans expanded the frontiers of Greek arithmetic and geometry, including work on triangular and polygonal numbers, the classification of odd and even numbers, the arithmetic, geometric and harmonic means, the nature of the irrational, and of course the Pythagorean Theorem.

Who among the Pythagoreans was responsible for which advances has been lost to time, and even the ascription of the Pythagorean Theorem to Pythagoras is more a matter of tradition than certainty. Theano comes down to us as the most eminent of the women Pythagoreans, but which mathematical areas she worked upon, it is at present impossible to say. But here is what we know …

Most sources say that Theano was the wife of Pythagoras and the daughter of Brontinus, but some hold that she was the wife of Brontinus and a gifted student of Pythagoras. The most detailed account, that of Iamblichos (*c*.245–*c*.325 CE) in his *Vita Pythagorica*, has her as Brontinus's wife, but the fragmentary mentions of her in Eusebios of Caesaerea (fourth century), Theodoretos of Kyrrha (fifth century), and Timaios of Tauromenion (third century BCE) all have her as Pythagoras's wife, while Diogenes Laertius (third century) says she could have been either man's wife. Modern scholars are divided on the issue, with some maintaining that the confusion has arisen because there were, in fact, two people named Theano whose lives got conflated with the passage of time.

Those who say she was married to Pythagoras claim that, after his death, the Pythagoreans were held together by her and her children Telauges, Myia and Mnesarchos, thereby creating the first link of continuity that would allow Pythagoreanism to exist as a virtually secret society for 1,000 years.

Lukianos of Samosata referred to her in the second century BCE as 'the daughter of Pythagorean wisdom'. Areios Didymos, one century later, asserted that she was the 'first Pythagorean woman to philosophise and write poetry', while Censorinus wrote some 400 years after that, placing her authority next to that of Aristotle on a question of natal periods. Our first hint of what she might have written doesn't emerge until Suda's *Lexikon* of the tenth century CE, some 1,500 years after her death. The titles he mentions as being written by Theano are: *Of Pythagoras*; *Of Virtue (for Hippodamos of Thurium)*; *Apophthegmata of a Pythagorean*; and *Advice for Women*.

Of the contents of the two Pythagorean volumes we can only speculate, but we do have three letters purported to have been written by Theano, two common anecdotes from her life, and three further letters of a more dubious provenance to get a sense perhaps of her style.

The three letters are missives of advice sent to friends, and thus contain no mathematics. What they do contain is a blunderbuss of Pythagorean life-counselling – reprimanding one friend for spoiling her children with luxuries when austerity and a love of the life of the mind are most likely to produce good children; advising another to treat her servants with kindness and consideration and avoid unnecessary luxury; and pushing a third to react with an air of calm to her husband's infidelity because men are essentially short-sighted sexual fools from whom not much is to be expected.

The two anecdotes are repeated constantly in the sources, which is endlessly aggravating. 'Hey, future generations, which would you rather have, accounts of the mathematical insights of the early Pythagorean women or this one story about Theano's elbow? I can't hear you, because you don't exist yet, so I'm going to assume the elbow thing and go ahead and throw the rest of this stuff on the fire. You're welcome.'

So, the elbow story.

Theano was walking along one day when her elbow came uncovered. Somebody commented that it was a beautiful elbow. She said, 'Yes, but not a public one!'

That is the elbow story.

The other anecdote is a similar two-line exchange. Somebody asks Theano how long after sex it takes a woman to become pure. She supposedly answered, 'With your husband, instantly, with somebody else, never.'

I have my doubts on those two – among the dubiously attributed letters by Theano and other women Pythagoreans there is an awful lot of, 'A woman's job is to please her husband', which doesn't seem to jibe with the daring communal intellectual spirit of early Pythagoreanism, but which *does* jibe perfectly with the Christian Platonism tasked with transmitting the heritage of ancient Greece, and which often transmuted it in the process.

At this far remove, with so few sources at our disposal, it is simply impossible to say what among the histories, whispers, letters and anecdotes counts as the 'real' Theano. What is true is that, long ago, at the dawn of the Western intellectual tradition, a group of men and women gathered, spurred on by love of intellectual exchange, and the association that they formed was so compelling it stretched over centuries and millennia to prod us forward with its example. Here was Pythagoras, and there Theano, philosophers both, the heirs of Egypt and parents of us all.

FURTHER READING: We are fortunate to have all the ancient fragments about Theano and the women Pythagoreans gathered in one easily obtainable source: *Theano: Briefe einer antiken Philosophin*. It is a Reclam edition, so it is very convenient for travel, and contains the original Greek and Latin next to the German translations. For more on the Pythagoreans and what we actually know about them, Penguin has a nice volume, *Early Greek Philosophy*, that includes sections on Pythagoras and the two schools emerging from him. On the mathematics side, Sir Thomas Heath's classic two-volume *A History of Greek Mathematics* (1921) is a through and through classic of mathematical history that belongs in every history of science collection.

Chapter 2

Hypatia of Alexandria: Philosopher, Mathematician, Political Casualty

By 400 CE, Alexandria was submerged in a sea of political-religious rivalries that drew even the most innocuous-seeming of scholars into its ravenous maw. For centuries, the intellectual capital of the world, boasting the largest storehouse of scientific and cultural information ever assembled, a succession of feuding prefects and patriarchs, pagans and neo-Platonists, employing mob violence as an all-too-regular form of expression for political grievances, had reduced the gleaming city to a nervous husk of its former glory, and at the twitching centre of that husk lay two of the greatest scholars in its long history: a father and a daughter.

The father's name was Theon, and the daughter was Hypatia (*c*.370–415). They were the inheritors of one of the most robust mathematical traditions in the history of the world, the terminus for a line stretching back seven centuries through Ptolemy, Diophantus, Apollonius and Euclid. Had they lived in better times, they might have been the originators of startling new mathematical theories. But they didn't. They lived under the rule of Theophilus and Cyril, two patriarchs who did not shudder before the use of violence to neutralise their religious and political opposition. According to Socrates Scholasticus, Theophilus was ruthless in scouring pagan activity from the city, possibly destroying the last volumes from the Great Library in the process, and Cyril employed his predecessor's Nitrian monks and armed mobs to terrify pagans, Jews and members of rival Christian sects alike when they challenged his authority.

For a person of conscience in that atmosphere, the preservation of the past in the face of an uncertain future was a high calling, and both Theon and Hypatia devoted themselves to preserving the most important mathematical concepts of the past so they might not be lost forever. Theon wrote definitive commentaries on Euclid and Ptolemy, the former of which was our primary source for Euclid's *Elements* until the nineteenth century. It was left to his daughter, then, to continue the tradition and attempt to capture the most recent developments in mathematics.

Most of what she accomplished has been lost to us, swallowed in the twisting vortex of persecutions, censorship and neglect that followed her violent death, but we at least know what she wrote about, and that was the conic theories of Apollonius and the algebraic theories of Diophantus. Conics include hyperbolas, parabolas and ellipses, and model everything from the path of an object in projectile motion to the orbits of planets and comets to the stored energy in a compressed spring. They had been investigated prior to Apollonius, but his expanded treatment of them, and in particular the addition of quasi-Cartesian reference frame elements, was the definitive statement of antiquity's geometric genius.

Diophantus, meanwhile, investigated methods for finding particular and general solutions to algebraic equations. The problem that we will see Julia Robinson become famous for cracking was a Diophantine equation, as is Fermat's Last Theorem. Diophantus was interested in equations of several variables for which only rational answers were allowed (though today Diophantine analysis only allows integer solutions). What possible rational values of a, b and c are there such that $a^2 + b^2 = c^2$? Is there a way to generally categorise all possible triplets of answers? This thinking, had it been followed through, would have allowed European number and algebraic theory to grow and flourish as its geometric thought had. As it was, that mathematical rebirth would have to wait a millennium, when Arabic algebraic techniques reinvigorated Western thought.

These then, and perhaps much more, were the subjects Hypatia wrote about. According to the few scant remnants we have, and second-hand accounts of her work, she made no original contributions to these fields, but contented herself with producing clear editions, which included worked-out examples that clarified the original authors' points and checked their results for a more general readership.

For Hypatia was, above all things, a teacher. Followers thronged to her dwelling to hear her talk about mathematics, astronomy and Neoplatonic philosophy. After the death of her father, she was one of the world's most prominent mathematicians, a woman who could speak of the most modern developments in science and mathematics and their connection with the great Greek philosophical tradition.

Theophilus had had a decent relationship with both Theon and Hypatia. He saw them as harmless neutrals whose Neoplatonism was (in its most abstract incarnation if not in the particular form practiced by Hypatia) highly compatible with emerging Christian philosophy (Augustine of Hippo, Hypatia's contemporary, would in fact earn himself a sainthood for his cunning if curious amalgam of Neoplatonism and Christianity). That is not to say Theophilus was a nice guy. His use of violence to destroy pagans he didn't find useful was brutal and complete. But he was at least willing to let Theon and Hypatia be.

Not so his successor, Cyril. Cyril seems to have believed that his arch-rival for pre-eminence in the city, the prefect Orestes, had formed an alliance with Hypatia to put himself in good graces with the city's remaining but significant pagan population, and that therefore, to enhance his own position and make Orestes more pliant to his future plans, Hypatia had to go. We don't know if he gave the order to eliminate her, but he stood to benefit from her demise (at least in the short term) and had not shied away from employing violence to solve political problems in the past (much of the tension between Cyril and Orestes lay in each's willingness to use their authority to foster politically motivated murders and riots), so it is at least likely that he was responsible for stirring up the antipathy to her that resulted in her untimely and gruesome end.

Sadly, Hypatia's death is the best documented part of her life. While we have to sift through scraps and stylistic theories to attempt to reproduce her living work, we have multiple, if not entirely consistent, sources for her grizzly finale. She was stopped by a Christian mob while riding through the streets in her carriage. They seized her, dragged her inside a nearby church, and beat her to death with roofing tiles before ripping her body apart, limb from limb, removing her eyes, and burning the hewn pieces of her body outside the church. They had, in a frenzy of blood, destroyed one of the few fragile connections their city had with its glorious mathematical past, and paved the way for its steady descent into a shadow of its former self.

Cyril was declared a saint in 1883 for his contributions to Christianity.

FURTHER READING: There is a good deal written about Hypatia, which is somewhat surprising given the absolute dearth of information we have about her. Most of it, however, is fiction, and most of the non-fiction is not in English. For the English speaker, an easily accessible source is Michael A.B. Deakin's *Hypatia of Alexandria: Mathematician and Martyr* (2007). It contains not only a biography but appendices about the mathematics Hypatia is thought to have studied and complete translations of all the original source material we have pertaining to her life and work.

Chapter 3

The Algebraist of Baghdad: Sutayta Al'Mahamali's Medieval Mathematics

It is a thousand years ago. Europe is a stumbling, superstition-addled giant, depleting its energies on visions of holy violence and shuddering at half memories of former greatness. It has lost its past, and despairs at its present, and if you are a person seeking to know the ways of nature, it is not the continent for you.

For you, a curious soul of the mid-tenth century, there is but one place you must go: Baghdad, the Islamic capital built by the legendary Abu Jafar al-Mansur, the House of Wisdom.

In this city, science rules. The Caliph sends his scholars to scour the globe for manuscripts bearing lost Greek wisdom. He invites delegations from India to explain their curious ten-digit number system, and dispatches his mathematicians on missions to measure the curvature of the Earth. Paper technology imported from Asia is pressed into service to fuel a bustling book industry that itself supports a small army of translators, commentators, poets and researchers, all rushing to feed a population and government drunk on the glories of the written word.

If you happen to find your way to this glimmering, learning-mad city, and if you have a mind for mathematics, then there is one particular call you must make: to the home of judge Abu Abdallah al-Hussein. Here you will find his daughter, Sutayta al-Mahamali (d. 987), as renowned for her legal mind as for her mathematical mastery; a woman of genius widely celebrated as such by her culture, praised for her abilities by three of the era's greatest historians, and today sadly reduced to the status of a historical footnote.

We know more about her father, son and grandson, who were all well-regarded judges and scholars, than we do about her, meaning we must piece her life and work together from a few scraps of information and a wealth of intellectual context if we are to know her at all. The 'hamal' in her name comes from the Arabic word 'to carry', pointing to her family's origins as carriers of goods and people. By her own time, however, her family was well established in scholarly circles, and Sutayta had the benefit of a string of learned teachers.

She studied Arabic literature, jurisprudence, the interpretation of sacred texts, and mathematics a full 200 years before Europe produced women of comparably broad education and fame in the form of Heloise of Argenteuil and Trota of Salerno. We know that she was widely consulted for her legal and mathematical insight, and that she solved problems of inheritance that imply an advanced knowledge of that era's hot new field: ALGEBRA.

Questions of inheritance, of how to correctly distribute the proceeds of an estate between people of varying relation to the deceased, were mathematical hornets' nests until they fell under the keen mind of Muhammad ibn Musa al-Khwarizmi (d. 850), the man who combined the Greek algebraic problem-solving methods and proof formalism of Diophantus with the base-10 numerical system of Indian scholars to create what we recognise as the forerunner of modern algebra. (The name algebra, in fact, comes from the term al-jabr that al-Khwarizmi coined for the operation of rewriting an expression to eliminate negative terms.) In his works, he applied his techniques to the complicated systems of equations that result when you try and mathematise a web of competing claims on an estate.

Sutayta, we are told, made original contributions to this field, and to the theory of arithmetic as well, which was truly coming into its own thanks to the adoption of a numerical system that wasn't aggressively hostile to computation. (To get an idea of how fortunate we are to live in the world of Indian/Arabic numerals employing base 10, just look up how awful it was trying to multiply with Roman numerals, or ponder for a while what it would be like to use the Babylonian base-60, which would require children to learn 1,711 times table products instead of the thirty-six required by our base-10. Something to use the next time your child complains about having to learn their nines.) Unfortunately, the exact solution types she contributed to mathematics have been lost in the intervening centuries. We know that she memorised the Koran, that she was praised for her virtue and modesty, but the historical record doesn't mention a single specific equation of her invention, in spite of the fact that we know that later mathematicians referenced her work.

To know her research at all, then, we must make a comparison to mathematicians of her time to get a sense of her areas of study. She lived 100 years after al-Khwarizmi and just after the time of Abu Kamil (850–930), the two titans of classical Arabic mathematics. Both of them were interested in categorising solutions to entire systems of equations, like al-Khwarizmi's breaking of all quadratics (equations that have a squared variable) into six solvable categories (or so he thought), or Abu Kamil's solutions to problems involving irrational quantities.

It is mentioned by her biographers that she was known not only for solving individual problems but for creating general solutions to *types* of problems, which would be a logical extension of the work of al-Khwarizmi and Abu Kamil. And, since her work both followed theirs and earned enough esteem to be mentioned by later sources, it is at least probable that she achieved a remarkable level of algebraic sophistication that opened groups of equations to Arabic mathematicians, extending beyond those solved by the great Abu Kamil, perhaps even including the solution of the cubic type equations that gripped the imaginations of her near successors Ibn al-Haytham (965–1040) and Omar Khayyam (1048–1130).

All this, however, is speculation based on interpolating from the work of those algebraists just before and just after her time, and from a few tantalising clues left by historians from the next century. With so little to go on, you might fairly wonder why we bother to include her at all in a history of women's contributions to mathematics – why talk about somebody we only know vaguely from reputation when there are so many current women of science whose work we are well familiar with?

Simply put, there is historical value in the question, *How was Sutayta even possible?* The relation between Islam and women's education is a deeply intricate tangle of general principles losing themselves in the messiness of local and ancient tradition. What scholars of the era are in general agreement upon is that Islam was the victim of its own early successes, and women more so than men. The phenomenal rate of expansion of the early Islamic Empire meant that, before they were entirely ready, the original Islamic faithful, including a number of powerful women, found themselves outnumbered in their own suddenly massive state.

The majority of the empire's population now consisted of far-flung tribes and peoples that clung to a more repressive and traditional pre-Islamic conception of a woman's place, and used their influence to retroactively write their beliefs into the core principles of the Islamic state. Once those beliefs were well integrated with current practice, Islamic historians, who considered pre-Islamic Arabic civilisation to be little more than a collection of barbarisms, rewrote the lineage of those beliefs to make them Islamic inventions rather than pre-Islamic holdovers.

Meanwhile, that edited version of the roots of Islamic practice was subjected to the mangle of Islam's roving capital, as Spain, Egypt and

Mesopotamia all tried their hand at melding its tenets with their dominant cultural practices. The result was a complicated spectrum of approaches to women that belies the monolithic Western narrative of veil and darkness.

That spectrum, at least for a time, had enough breadth to include a mathematician like Sutayta, a female poet of erotica like Wallada, and a wealthy political powerhouse like Khayzuran. The constriction of that breadth, and the decline of the House of Wisdom, would come in time, but the knowledge that, once upon an empire, there was a learned city wherein walked a woman who wove law and algebra with equal ease and to general acclaim can give us hope that what once was, might be again, and perhaps already is.

FURTHER READING: Tesneem Alkiek has made a brief YouTube video about Sutayta that is a good starting point for the curious. For more on the history of Arabic mathematics and the early history of algebra, the best book is Jacques Sesiano's *An Introduction to the History of Algebra*, which includes original Greek and Arabic texts in a valuable appendix. Jonathan Lyons's *The House of Wisdom* contains a lovely chapter on the intellectual atmosphere of late tenth-century Baghdad, and Ehsan Masood's *Science and Islam: A History* has as good a history of Arabic mathematics as you'll get outside of Sesiano's harder-to-find book.

For the history of women in the early Islamic Empire, Wiebke Walther's *Women in Islam from Medieval to Modern Times* has a good account of the complex evolution of women's position throughout the empire, and some engaging portraits of women active in politics and writing, though not so much on science. Finally, *Women in Iran: From the Rise of Islam to 1800*, edited by Guity Nashat and Lois Beck, has a chapter by Richard Bulliet on how and why different types of women were covered by the great Islamic biographers, which sheds light on the difficulty of knowing more about Sutayta than we do.

Chapter 4

Genius Overcome: The Destruction of Catherine de Parthenay

The life of Catherine de Parthenay (1554–1631) was dominated by national religious conflict and her decision to take a pivotal role of resistance at the all-consuming centre of that maelstrom. The war between the Catholics and Protestants in France during the sixteenth and seventeenth centuries robbed Parthenay in turn of her home, her health, her children and her freedom, leaving one of France's most brilliant women, at the end of her long life, wandering the ruins of her former estate with her only surviving daughter, ruminating endlessly but without self-pity on the heavy cost of her unwavering religious loyalty.

For somebody whose life ended in such omnipresent gloom, Parthenay's life began in the full bloom of promise. Her father, Jean de Parthenay, was the lord of Soubise and Mouchamps, and a courtier known for moral integrity and intelligence, while her mother, Antoinette Buchard d'Aubeterre, was a maiden of honour at the court of the future Queen of France, Eleanor of Austria (1498–1558), also known as an individual of culture and brilliance, and a devotee of the religious reform movement bubbling to life in early sixteenth-century France. The pair were married in May 1553, and Catherine was born in March 1554.

Whereas many daughters of families with close connections to the royal court grew up with an education centred around dress, etiquette and ostentation, the Parthenays, influenced by their Protestant principles, raised Catherine to honour humility, intelligence, simplicity and piety, and so it was that the young girl, while lacking almost entirely friends of her age to play with, or any toys beyond simple rigid dolls, or the fine dresses and garish makeup that were the common currency of her peers who frequented the notoriously loose-moralled court of Francis I, was compensated for those deprivations with access to all manner of books, and to high-quality teachers willing to explain their contents.

The greatest and most influential of those teachers was doubtlessly François Viète (1540–1603), who entered the Parthenays' service in 1564, two years before the death of Jean. Viète would serve as Catherine's teacher until 1570, during which time he taught her mathematics, astronomy, astrology, Greek and Latin, while in turn learning about Calvinism from the family. He found Catherine a brilliant student, who learned new languages as easily as she did novel mathematical concepts, and as she soon outstripped the books she had in her possession, Viète took the step of writing treatises for her to study, based on his own advanced mathematical explorations, including the use of decimals (which had been employed by Muslim mathematicians in the fourteenth century, but which wouldn't become common in Europe until Simon Stevins's book *De Thiende* in 1585), and his theory about planetary orbits being elliptical (another scientific instinct that wouldn't become common in scientific circles until Kepler's planetary laws were published forty years later, in 1609).

Poet, linguist, mathematician, Catherine de Parthenay was already an intellectual juggernaut by 14 years of age, and had she lived in a more modern era she would probably have had at least another half decade to develop those attributes. As it was, she lived in the mid-sixteenth century, and as such was, upon turning 13, of an age to enter the marriage market. She was originally intended to marry Gaspard de Coligny, the 15-year-old son of one of France's most distinguished Huguenot families and a close friend of the family, but his death at the hands of the plague meant Antoinette had to reshuffle the dynastic deck and find a new suitable match.

She fatefully decided upon Charles de Quellenac (1548–1572), who was far more of an unknown quantity to the Parthenays, and the pair married in 1568, only to discover that they were not particularly interested in each other. Charles was devoted to serving the Protestant cause through force of arms and, in his absence, Catherine returned to the ways of her childhood, so that on those occasions when they did manage to be in each other's company, there existed a reigning awkwardness that only seemed to get worse with the passage of time. Antoinette felt that she had made a profound mistake in arranging the match, and pushed her daughter to obtain a divorce. Catherine, though not rapturously happy in her marriage, was not able to completely commit to her mother's divorce strategy, however, and it was not until 1572, when Charles was killed along with tens of thousands of other French Protestants during the St Bartholomew's Day Massacre, that the generally dissatisfying wrong turn in her life found its resolution.

Catherine was in La Rochelle, not Paris, during the Massacre and so avoided the immediate aftermath of the carnage, but as La Rochelle was a Protestant stronghold, it was inevitable that it would soon fall under siege. Brought up on overwrought tales of the heroism and justness of the

Protestant cause, she found herself wanting to aid its fight against oppression with every fibre of her 19-year-old being but, lacking the ability to join the military effort against the forces of Catholicism, Catherine instead employed her brain, writing a play, *Holopherne*, to dramatise the Protestant cause, and particularly the brave resistance of La Rochelle.

These were years of almost constant religious war in France, with each peace only giving way to new conflict as new centres of resistance and combinations of state power waxed and waned. To Catherine, these were years when Protestantism showed its indomitable spirit, but to her mother, they were perhaps more significantly the chaotic backdrop behind the real story, which was finding a new husband for her daughter, one of good family, stout character, firm Protestant convictions, and who was capable of forming an actual emotional attachment to her daughter. The perfect solution soon presented itself in the form of René II de Rohan (1550–1586), Lord of Blain, Protestant soldier par excellence, friend and supporter of Francois Viète (who still kept in contact with Catherine as he was working his way towards his 1591 magnum opus that proposed many aspects of modern algebra), and a charismatic leader whose services were equally sought after by royalist and reformer alike.

The couple were married in 1575 and this time all was well. Catherine was enchanted by Blain, with its combination of deep family archives and sprawling experimental gardens, and found in René an individual as devoted as she was to the Protestant cause but who was still capable of being equally devoted to her and their growing family. A daughter, Henriette, was born in 1577, and another, Catherine, in 1578, with a son, Henri, following in 1579. Catherine's next child died shortly after childbirth (1581) and the following was stillborn (1582), but in 1583 a second son was born, Benjamin, with 1584 seeing the birth of their last child, the long-suffering Anne.

From 1578 to 1585, René kept to his resolution to stay at home with his family and not risk himself, and thereby their future, in battle. These were happy times for Catherine, probably the last happy times she would ever truly know, as her husband stayed safe beside her, and her children grew under her care and education to honourably reflect the values and beliefs she held dear (including her love of mathematics, to which she returned during this brief idyll). Finally, however, René could no longer resist the call to action, and rejoined the war effort to fight for the cause of Henri of Navarre in 1585. He soon found himself in La Rochelle during an outbreak of plague which claimed his life in April 1586, leaving Catherine a widow once again, with the responsibility of not only raising five children on her own, but of administering multiple hotbeds of Protestantism in the middle of a tumultuous civil war that would not see its first era of semi-calm until the promulgation of the Edict of Nantes in 1598.

The coming decades saw Catholic armies raze any property associated with her name, destroying Blain with a particular vengeance that extended even to the family archives. The looting of her property and destruction of her land left Catherine in constantly dire financial straits, while her role in one of the great conflicts of the French religious civil wars culminated in the breaking of her health and the loss of her freedom. In 1627, she was in La Rochelle when the troops of Louis XIII placed it under siege. While her son Henri attempted to break the siege by force of arms from without, she maintained the spirit of resistance from within the city, encouraging the citizens not to give in to royal pressure, to repeat their great stand of 1573. It was heroic and inspirational advice, surely, but perhaps not ultimately in the best interests of La Rochelle, which over the course of its fourteen-month ideologically sustained resistance saw its population dwindle by 80 per cent, from 27,000 to just around 5,000, at the hands of disease and starvation.

The defence of La Rochelle ended in failure, and for their role in encouraging it, Catherine, aged 74, and her daughter Anne, aged 44, were imprisoned, remaining in jail from November 1628 to July 1629. Returning to Soubise after their release from jail, Catherine and Anne found a world stripped of its meaning and memory. Their lands razed, their greatest friends from the decades-long Protestant struggle dead from disease and battle, their finances devastated, and their family broken (Henriette and the younger Catherine had died of health complications, and Henri and Benjamin were both exiles thanks to their role as Protestant soldiers), the pair haunted their ancestral lands, lingering over old wounds, and marking the steady dwindling of their threadbare social circle. Catherine's health, already in a perilous state after fourteen months of steady starvation, and eight months of prison, deteriorated steadily, until she finally breathed her last on 26 October 1631.

Catherine's intellectual journey during these years was one shared by many educated women of her era. Undoubtedly one of her age's great minds, she absorbed with alacrity the newest developments in science and mathematics while keeping herself grounded in classical traditions and the reverence of dynastic history. She and René were both firm supporters of Viète's work, which support he acknowledged by dedicating his 1591 *Analyse mathématique restaurée* to Catherine. During the rare tranquil moments in her life, and particularly during the era when she was educating her children, she returned to her love of mathematics, but the pace of political and religious events, and her parents' example of steadfast devotion to a religious cause, pushed her ever away from her purely personal studies and towards the hungry centre of French theological politics, which would ultimately devour her and her family utterly. She published no mathematical volumes, and produced no new theories. Though she had a mind equal to those tasks, she lived in a world set firmly against their realisation, leaving us with the sad conclusion that genius does not always, after all, sweep every difficulty before it, and that a culture that lionises inflexible obedience to a cause often ends by consuming its greatest minds.

FURTHER READING: The book here is Nicole Vray's *Catherine de Parthenay, duchesse de Rohan* (1998), which as far as I know has not been translated but which I found pretty easy going with my old university French, thanks to Vray's engaging style and gift for narration. In English, your choices are far less robust, consisting primarily of Parthenay's section in the essential core book for all Women in Science enthusiasts, Marilyn Ogilvie and Joy Harvey's two-volume *Biographical Dictionary of Women in Science*.

Chapter 5

Brief Portraits: Antiquity to the Early Modern Era

Occello or Eccello of Lucania (Fifth/Fourth century BCE)

Seventeenth-century French scholar Gilles Menage makes mention of an Eccello or Occello from the southern Italian region of Lucania as having been a member of the Pythagorean order, but beyond that we know nothing about her work or life.

Pandrosion (Fourth century CE)

According to Greek mathematician Pappus of Alexandria, Pandrosion was a fourth-century mathematician known for a three-dimensional recursive method of calculating the cube root of a number. After a brief period of uncertainty as to Pandrosion's actual gender, most historians today have come down on the side that Pandrosion was, in fact, a woman mathematician pre-dating Hypatia of Alexandria.

Maria di Novella (Fourteenth century)

The Bolognese Maria di Novella was one of the early representatives of a type of figure that became common in Renaissance Italy and beyond: the brilliant home-educated woman polymath, of whom the eighteenth-century Maria Agnesi is perhaps the most famous example. Di Novella was the daughter of John Andreas, a professor in Bologna, who educated his daughter in natural philosophy and mathematics, and who is rumoured to have sent her to lecture in his stead whenever he was ill.

Theodora Dante (1498–1573)

Dante was born in Perugia, in the dead centre of the Italian peninsula and, in the tradition of Maria di Novella, attained renown as both a miniatures artist in the manner of High Renaissance master (and teacher of Raphael) Pietro Perugino, and as a mathematician, in both of which disciplines she was privately taught.

Elena Lucretia Cornaro-Piscopia (1646–1684)

The first woman to receive a doctorate degree from a university, the Venice-born Elena Cornaro-Piscopia was the illegitimate child of Gianbattista (or Gio Baptista) Cornaro-Piscopia and Zanetta Boni, a peasant woman. The couple later married, in 1654. Her father, who was among the leading members of the Venetian nobility, ensured that she had a broad-based education that included, by her teenage years, the mastery of seven languages, four musical instruments, and a deep knowledge of physics, mathematics, astronomy, philosophy and theology. Her 1669 translation of the early sixteenth-century monk Lanspergius's *Colloquy of Christ* ran through five editions in three years.

In 1678, she underwent a public examination whereby she was given random passages of Aristotle to translate and explain in the presence of Venice's leading lights and key figures from the universities of Bologna, Rome and Naples. Succeeding in this, she was awarded a doctorate degree from the University of Padua, the first ever awarded to a woman by a university. The desire of other fathers to have their daughters awarded similar degrees caused the universities of Italy to roll back their degree policies, and it was not until 1732 that Laura Bassi became the second woman to hold a university degree. Elena subsequently taught mathematics at the University of Padua, though her writings from this time have not survived. She died of tuberculosis on 26 July 1684.

Chapter 6

Sex, Cards and Calculus: A Day with Émilie du Châtelet

In popular mythology, the 1687 publication of Newton's *Principia* was *the* culminating moment when one human told the world how the universe works, threw up his arms in triumph and then received the adulation that was his due. Of course, it worked nothing like that, and while England was quick to lionise his intellectual achievement, it took half a century for many of his ideas to catch on in continental Europe. For reasons at best understandable and at worst downright xenophobic, the scientific community of the mainland spent those fifty years turning itself inside out trying to do deference to every half-expressed notion of Descartes or Leibniz while steadily and insistently denigrating Newton.

Until Émilie du Châtelet.

She was raised with a better than average education for a woman of her era, which is to say that she had a good grounding in the Classics, and a particular gift for languages (she would eventually learn English, Latin, French, German and Flemish), but was not given an education in science or mathematics until after she married and had the wherewithal to hire her own private tutors.

And there began one of the truly amazing stories in the canon of intellectual history. Starting from virtual scratch at the age of 28, she made of herself one of the leading European mathematicians by the time of her death a mere fifteen years later. Beginning from the basic concepts of algebra, she devoted a dozen hours a day to the improvement of her mind, sometimes plunging her arms into icy water to keep herself awake deep into the morning hours in pursuit of a particularly thorny problem.

Had she undertaken this regime on her own, a single woman of means pursuing her passion, it would have been impressive enough, but her life was bursting with other activity besides. Married off at 19 to the Marquis du Châtelet, an older, laissez-faire sort of gentleman, she had the responsibilities of running the household and raising the children to attend to, as well as the mammoth set of complications that arose when she decided to take on as a lover the greatest but most trouble-prone author on the continent, Voltaire. Together, the pair set up a scientist's paradise in Cirey with the blessing of the indulgent marquis, and they spent their days in experimentation and Newtonian studies, eventually jointly writing a staggeringly popular book that attempted to explain Newton's ideas in popular language.

But Émilie wasn't done. Though Voltaire was content to remain a partisan Newtonian, Émilie knew that the full truth, if it was ever going to be found, would lie in an unbiased and impartial review of all the ideas currently available, to which end she researched and wrote her great *Institutions de physique* (1740), a review of the history of physics down to her own day, with equal weight given to the insights of Descartes, Leibniz and Newton. While praising Newton's mathematical rigour, she also noticed that his reliance upon the hypothesis of absolute space was potentially problematic (as it in fact turned out to be, though it took another 200 years for us to realise it). At the same time, she recognised that Leibniz's notion of *vis viva* was something independent of Cartesian momentum and advocated for a more thorough investigation of its properties (which, thanks in part to her, ultimately happened, and today we call it Kinetic Energy).

Still, though, the resistance to Newtonian gravitation remained, so she undertook the towering effort of her later life. She resolved not only to translate all 521 pages of the *Principia* into French but also to add an extensive commentary that included all of the data gathered in the intervening half century that supported Newton's theories. It was a massive undertaking that only somebody with a full and decisive grasp of both mathematics and the state of contemporary physics could have even contemplated. Luckily, she had such a grasp and, in spite of starting a second love affair with a younger officer poet (her relationship with Voltaire at the time had settled down into something pleasantly platonic, so she wasn't in fact double cheating on her husband, who didn't mind in any case) and becoming pregnant by him at the age of 43, she managed to finish the task just before giving birth to a daughter on 4 September 1749.

Émilie du Châtelet died six days later and ten years after that her translation and commentaries of Newton's *Principia* were finally published, a translation that remains the standard in France to this day. Its deep synthesis of Leibnizian methods and notation to enhance Newton's original examples presents a broadness of perspective that was without peer in its age, and one cannot help but grieve over the loss of what she might have accomplished had that final pregnancy not ended her life prematurely.

Or perhaps one can help it. She lived a marvellous and full life far beyond the expectations of her time. She did what she loved, whether it was staying up all night working through formulas or chatting with Voltaire for hours about materialism and the inaccuracies of the New Testament, or throwing herself into the arms of impossible romance or losing immense amounts of money at the gambling table. And she inspired a wave of women to follow her example, including the worshipful and brilliant Francoise de Graffigny, who would go on to write one of the most popular novels of the century. Her philosophy of personal joy and improvement, as she wrote it in her 'Discours sur le Bonheur' knew no artificial borders of tradition and is, today, in our era of trading grand passions for a flood of responsible micro-happiness units, perhaps worthy of turning our weary heads towards once again.

FURTHER READING: Since the 1970s, Émilie has been making a steady comeback in the public consciousness. If you can find a copy of Lynn M. Osen's 1974 *Women of Mathematics*, the portrait of Émilie in it is brief but sparkling and poignant. More recently, Émilie makes up half the material of Robyn Arianrhod's 2011 *Seduced by Logic: Émilie du Châtelet, Mary Somerville and the Newtonian Revolution*. I'm not a great fan of the title, but the book within is wonderful, containing not only an account of these two astounding scientists but a thorough treatment of the development of Newtonian thought in the eighteenth century with some smashing mathematical appendices.

For the fiction reader, Adrienne de Montchevreuil in J. Gregory Keyes's *Age of Unreason* tetralogy is basically Émilie, except with the ability to command ethereal spirit creatures in addition to her wicked awesome mathematical abilities. So, yeah, that's worth a read.

Chapter 7

The Curve Who Became a Witch: The Geometric Calculus of Maria Gaetana Agnesi

If any century favourably understood the manic blend of child shaming and twisted pride that is the typical *Toddlers and Tiaras* pageant parent, it was the eighteenth. Child prodigies were in, and if you were aching to claw your way into the ranks of the minor nobility, your precocious son or daughter was your meal ticket. Some decades before Leopold Mozart dragged young Wolfgang to any prince or archbishop who had half a chance of offering a decent appointment, a Milanese girl with a genius intellect was made the centre of an ongoing academic circus routine by her status-hunting father.

She would go on to write one of the first comprehensive calculus textbooks in Italian, and then suddenly forsake all scientific study to devote herself completely to the well-being of the poor and elderly. She was Maria Gaetana Agnesi, remembered today only for a curve, the *Witch of Agnesi*, which shows up in the margins of introductory calculus texts from time to time, and which she didn't actually discover. In her own age, though, she was recognised as an intellectual wonder of the world. Born in 1718 in Milan to a merchant family with big dreams of gaining entrance to the nobility, she and her sister were from the outset given an intense education by a series of top-notch tutors. Maria showed an early genius for languages, philosophy, science and mathematics, while her sister attained renown for her musical compositions.

Their father, Pietro, a spendthrift of the most abject order, saw in them his ticket to greatness. His plan was two-fold: (1) Spend money as ostentatiously as possible to get the nobles to respect him, and (2) Arrange a series of entertainments for the religious and noble orders with his children as the stars. These were the famed Agnesi *conversazioni*, in which the young Maria would be seated at the centre of a circle of onlookers, and instructed to answer any question about any topic that might come to their minds, in any language they chose. She knew seven languages by her tenth birthday, could compose academically rigorous defences to proffered theses on the fly, and hated every moment of it.

She was reserved by nature, and the strain of her performances plunged her into a serious illness at the age of 11. While in the depths of her sickness, she was offered religious instruction by priests from the Theatine order whose approval her father was seeking. They took the overworked, sick girl and threw on her shoulders an ascetic regime that taught the denigration of worldly emotions and a selfless devotion to pure intellectual effort and tireless charity.

She got better, and credited their regime with her recovery. She renounced her earlier philosophical speculation and scientific curiosity, and plunged into the realms of pure mathematics, social work and theology. It was a programme well fitted to the principles of the Milanese Catholic Enlightenment – an odd amalgam that attempted to hold together a profound admiration for the accomplishments of Galilean and Newtonian science with a fundamental belief in the basic correctness of the Catholic faith. They sought to live a life of unrestrained intellectual curiosity augmented by a service-centred approach to religious life.

The Catholic Enlightenment was full of interesting and well-meaning ideas that were destined to satisfy nobody. Agnesi divided her time between working at a care hospital for the elderly and writing a two-volume textbook that sought to synthesise and explain, for the first time in Italian, the insights of analytic geometry and calculus. The work, *Instituzioni analitiche*, was finally published in 1748 and made her an academic star, resulting in an invitation to join the faculty at the famous University of Bologna, where Laura Bassi also served as a professor of physics.

The book itself is something of a curiosity. It was Agnesi's opinion that the application of calculus to physical problems was profoundly uninteresting because it was merely worldly, and so her book intentionally leaves out many of the advances attained by Continental mathematicians of the Leibnizian tradition in favour of a return to Cartesian and Newtonian geometric arguments. The English and Catholics, predictably, loved it, while others viewed it as a noble relic, the last and best effort of a tradition that decidedly did not have the wind at its back. It was the summit of an approach to mathematics that would soon be swallowed by the analytic power of Lagrange and Euler, and perhaps the most important scientific work to come out of the Italian Catholic Enlightenment.

It was also the last thing Agnesi wrote publicly about mathematics. The book done, she retired completely into her charity and religious work. For the remaining fifty years of her life, she gave comfort to the mentally ill and elderly as the director of the Pio Albergo Trivulzio from 1771, and hunted the streets for children to be brought to her catechism class. The woman who had been regarded as one of the greatest minds of Italy, consulted from every corner of Europe for her advice on matters mathematical and philosophical, died unremarked in 1799.

FURTHER READING: The major book available on Agnesi is *The World of Maria Gaetana Agnesi, Mathematician of God*, by Massimo Mazzotti. It is under-long, over-priced, and in general, I don't like it. It represents that academic tradition wherein the mad desire to win fame by coining a trendy theory gets in the way of equitably assessing the material at hand. There are some good and interesting slices of history we would not have seen except through Mazzotti's research, but there is always a lurking theory-born torsion twisting the sources in a way that makes you not quite trust the conclusions, such as they are. Unless you are particularly curious about the secondary and tertiary figures who had a hand in the Milanese Catholic Enlightenment, you're probably better off looking at some of the briefer lives of Agnesi, like that in Lynn Osen's *Women in Mathematics*.

Chapter 8

Primal Screams: Sophie Germain's Mathematical Labours

It is a well-known fact of humanity that the chances of a group of people electing to do something decent and necessary is inversely proportional to the number of people in that group. We enshrine and attempt to forgive that principle under the banner of Institutional Inertia, but the fact remains that very decent individuals have, when given the power to act in concert, caused a great amount of pain in the history of science, with few examples as consistently pathetic as the ignored pleas of mathematician Sophie Germain (1776–1831) to have somebody, anybody, from Parisian academic circles acknowledge her theories publicly, or even provide feedback privately.

Germain's life spanned France's most turbulent political and intellectual years. Regicide, Revolution, Empire, Restoration, the Hundred Days, another Restoration, and another revolution, all grinding past each other in dizzying succession while there gathered in Paris an All-Star roster of mathematical geniuses. Lagrange, Laplace, Fourier, Legendre, Monge, Cauchy, Abel, Galois, Dirichlet, Poisson – you cannot read through ten pages of a calculus or number theory text without running into something named after one of them, and they were all in the same city, Germain's Paris.

We know precious little of her childhood except that, when revolution came in 1789, the young girl buried herself in the books of her father's study, teaching herself mathematics and science while the old order crumbled just outside her window, learning enough Latin somewhere along the way to allow her to devour and understand central mathematical texts like Gauss's landmark *Disquisitiones arithmeticae*.

There was a limit to how far books and books alone could take her, however. To learn more, she needed access to the minds of the people who were pushing out the boundaries of mathematics at such a reckless pace. As a woman, she couldn't attend their lectures, but she *could* write to them.

Worrying that they wouldn't take a woman correspondent seriously, Germain pretended to be a Monsieur Leblanc and wrote to both Lagrange and Gauss, asking questions and forwarding mathematical proposals that favourably impressed them both. Her mind stood out even amongst her generation of mathematical geniuses, and when they eventually found out her secret, that she was a self-taught woman, their admiration only increased.

Her first area of interest was Number Theory, and in particular the problem we now know as Fermat's Last Theorem, first proposed in the seventeenth century and only solved in the late twentieth. This theorem says that you cannot find natural numbers x, y and z such that $x^n + y^n = z^n$ for $n > 2$. For $n=2$ there are all sorts of natural numbers that work. You learn them in secondary school as the Pythagorean triplets: (3,4,5), (5,12,13), and so forth. But move just one power higher, Fermat theorised, and you will not be able to find a single set of triplets no matter how long or how hard you search.

By Sophie Germain's time, the theorem had been proven for $n=3$ and $n=4$, but a full solution for *any* value of n was proving elusive to the greatest mathematical minds of the age. Germain, of course, didn't solve it either, but her approach to the problem, and the ideas that she developed in seeing it through, represent important advances in not only the solution of Fermat's Theorem, but in the history of number theory. As such, her strategy deserves a bit of a closer look. So, nerd hats on, everybody!

Having solved the $n=4$ case, all that was really left was to show that the Theorem holds for any prime value of *n*. Why? Well, any natural number n greater than 2 must either be divisible by 4 or have an odd prime as a factor (think on that a while – it is a great little fact). For the former case, that means that we can rewrite $x^n + y^n = z^n$ as $x^{4k} + y^{4k} = z^{4k}$ and therefore as $(x^k)^4 + (y^k)^4 = (z^k)^4$, which is a form of the $n=4$ case which had already been proven to have no natural number solutions. For the latter case, if *n* has an odd prime factor, we'll call it *p*, such that $n = kp$, we can rewrite the equation as $x^{kp} + y^{kp} = z^{kp}$, and therefore as $(x^k)^p + (y^k)^p = (z^k)^p$, meaning that, really, the question we're asking is whether natural number solutions exist when the power is prime.

Enter Germain. She had an exquisitely beautiful realisation that picked up on Gauss's use of modulo notation in the *Disquisitiones*. He had introduced a new type of congruence based on remainders whereby a and b are said to be 'congruent modulo *n*' if they both have the same remainder when divided by *n*. For example, 42 and 7 are both congruent modulo 5, because both have a remainder of 2 when divided by 5. Germain took that idea and used it to construct a set of 'auxiliary primes' out of each prime power p to be investigated.

To construct the auxiliary primes for a particular p, she looked at all the remainders when you divide x^p by $2Np+1$, where *N* is any natural number.

If that set of remainders contained no consecutive numbers, she realised, then $2Np+1$ was an auxiliary prime for p and, most importantly, must divide either x, y or z. To take the simplest non-proven case, that of $p=5$ and $N=1$, we get $2Np+1 = 11$. When we repeatedly divide 1^5 through 10^5 by 11 (we stop at 10^5 because were we to go on we would hit 11^5 which is, of course, evenly divisible by 11) the remainders we get are 1 and 10. Those aren't consecutive, so 11 is an auxiliary prime of $p=5$, and therefore, if there is a triplet (x, y, z) that solved $x^5 + y^5 = z^5$, one of those numbers must be divisible by 11.

Germain's plan was to next prove that, for any prime p, there are an infinite number of auxiliary primes, and therefore that x, y and z must have an infinite number of factors, which is clearly impossible. Thus no (x, y, z) triplet would exist for any prime power p, and since the primes are all that was needed to prove the problem generally, Fermat's Last Theorem would have been put to rest. Unfortunately, proving that there are an infinite number of auxiliary primes was not something Germain, or anybody in her era, was equipped to do. The most she could do was to find enough auxiliary primes to show that, if a triplet existed, it must be insanely large, which seemed unlikely, but unlikely doesn't mean impossible.

She didn't prove Fermat's Last Theorem, but she developed new conditions for its solution, which broke the theorem into two sub-cases that later mathematicians could more efficiently whittle away at. In addition, she was the first mathematician to credibly forward a method that would attack the whole theorem at once, rather than just picking off single values of n here and there. In 1819, she sent her work to Gauss, hoping that the great reigning genius of number theory would advise her on how to proceed, but though Gauss had been friendly and encouraging when she had first written to him back in 1805, he did not, as far as we know, ever respond to Germain's questions. Lacking outside help to push the problem through, she did not publish her results herself, and indeed it was only through word of mouth, and a footnote in one of Legendre's books, that her significant contributions were known in France at all.

The refusal to publish makes sense when you look at it from the perspective of her other great contribution to mathematics: her differential equation for the elastic deformation of a disc. In 1808, a scientist and showman named Ernst Chladni came to Paris to show off his newest apparatus, a set of circular discs coated in a fine layer of dust that, when rubbed on the edge with a violin bow, formed distinctive patterns. The phenomenon fascinated everybody from society matrons to the Emperor Napoleon himself. In 1809, the Institut de France offered a prize to anybody who could mathematically explain the sand patterns that emerged from the vibration of the disc. Germain, who had studied the literature on vibrating bodies, decided to give it a shot and, indeed, hers was the only entry the Institut received, one which demonstrated a novel approach that made up for the errors in calculation that inevitably accompanied a problem so complicated.

The errors, however, were enough for the committee to not award her the prize. They announced a new competition with the same theme. She wrote up a new and improved version of her theory and, again, hers was the only submission. This time she got an honourable mention, but no prize. The committee then re-re-issued the challenge, and *yet again* Germain sent in her theory with further refinements and *this* time the prize was finally hers. In 1816, she became the first woman to receive a national prize for mathematics, but the victory was blunted by an unusually subdued and back-handed prize announcement which said, in effect, 'Yeah, it's an answer. It's okay, we guess. I mean, we'll give you the prize, if we *have* to.'

Germain understandably wanted to know just precisely what it was that the committee objected to in her work. She sent in a request for clarification, which the members individually promised quick action on and, as a group, entirely ignored. Nobody would tell her what she had done wrong to receive such lukewarm praise surrounding her prize. About every two years thereafter, she rewrote her results and sent them to the Institut to be discussed and entered into the archives and, each time, with much individual assurance that it would be attended to right away, her papers were stuffed in a dark corner and forgotten. Nobody would tell her what was wrong or how she could improve, a situation bottomlessly irritating for somebody who wanted to get at the truth.

The habitual snubs of the academic community were oddly paired with her celebrity in the wider world. Winning the prize made her famous, somebody to know and to be seen to know. On a person-to-person level, her friendships were warm and encouraging. Fourier, Gauss, Lagrange and Legendre were all friendly to her (though Poisson was a consummate ass who attempted to have her contributions written out of history). And yet, the institutions they largely ran repeatedly refused to officially evaluate and comment upon her work, letting it die unremarked or, at best, stuffed away in a footnote.

By the late 1820s, Germain had to curtail her mathematical activity in the face of a savage cancer that caused her too much pain to concentrate. She lived long enough to see the revolution of 1830, and also long enough to see the rise of a new generation of mathematicians: the wild and moody Galois, the tragic Abel, and the talented Dirichlet. Germain's body died before her curiosity had run out, and that is either quite tragic or extraordinarily beautiful, and most probably both.

FURTHER READING: Dora E. Musielak has written not only a fictional account of Sophie Germain but a mathematically intense biography of her, *Prime Mystery: The Life and Mathematics of Sophie Germain* (2015). It has a significant number of typos, but doesn't back down from the details of her work, which is refreshing. It is a very necessary counterpoint to the Germain bashing that has gone on for so long, and that as of the writing of this piece is still in evidence even on her Wikipedia page, so by all means check it out.

Chapter 9

Mary Somerville: Saviour of British Mathematics

In the 1750s, when France was foundering scientifically in the Cartesian shallows, it took Émilie du Châtelet's French translation of Newton's *Principia* to reinvigorate Continental physical science. Then it was England's turn to toss itself headlong into the longest period of scientific stagnation it has ever known. After the age of Newton, Harvey, Halley, Boyle, Hooke and Wren, there stretched an agonising century of nationalistic puttering. If England were to regain its mathematical groove, somebody would have to do for the British Isles what du Châtelet did for France.

Fortunately, somebody did. And her name was Mary Somerville.

She was born in 1780, the daughter of a chronically absent vice admiral of great renown but little fortune, and an indulgent mother who let her run wild. Her youth was an extended adventurous ramble through nature in the best Disney tradition, making friends with the birds of the forest until they ate crumbs from her mouth and affronting the sense of propriety of rich relatives. In her memoirs, she describes with picturesque whimsy the already vanishing colours of her Highlands youth:

> Licensed beggars, called 'gaberlunzie men', were still common. They wore a blue coat, with a tin badge, and wandered about the country, knew all that was going on, and were always welcome at the farm-houses, where the gude wife liked to have a crack (gossip) with the blue coat, and, in return for his news, gave him dinner or supper, as might be … There was another species of beggar, of yet higher antiquity. If a man were a cripple, and poor, his relations put him in a hand-barrow, and wheeled him to their next neighbour's door, and left him there. Some one came out, gave him oat-cake or peasemeal bannock, and then wheeled him to the next door; and in this way, going from house to house, he obtained a fair livelihood.

At the age of 11, her parents made a half-hearted attempt to civilise her, packing her off to a boarding school which featured, as part of its progressive programme, an iron chassis that all children were required to wear to correct their posture. It forced the shoulder blades back until they touched, and featured an extra iron loop that pushed the chin back to a proper position. It was agony added on to the slow burn of the school's uninspiring curriculum.

Mary returned from the school after a year, having apparently learned nothing whatsoever, and the experiment was mercifully ended. It wasn't that Mary's mind was dull, rather it was her fate to have it constantly entrusted to guardians with no notion whatsoever of what to do with it. On her own she learned Latin and piano and painting, and was distinguished in the pursuit of all three, but mathematics, which was to become the central love of her life, she knew not a wisp of until the age of 15, when she successfully engaged her brother's tutor in the task of procuring for her a copy of Euclid's *Elements* and a text on Algebra, it being socially impossible for a young lady to walk into a shop and purchase such items for herself without being the scandal of the town.

No thanks to her family, she finally had substantive, nourishing material to feed her brain, and she devoured it whole, staying up through the night studying her Euclid until the servants found her out and reported her late-night studies to her parents, who promptly forbid her the use of candles in an attempt to curtail her unfashionable obsession. Undaunted, Mary lay awake in bed, going over Euclid's proofs from memory ... which her parents also somehow found out about, and scolded her roundly for. Clearly, if her mind was to soar, it could not do so at home.

This suited her family just fine since – as readers of Regency fiction know all too well – the purpose of every early nineteenth-century British female was a marriage that would relieve her parents of the burden of caring for her. Mary was handed off in 1804 to a (and I use the term loosely) man, by the name of Samuel Greig who looked with scorn on the intellectual capacity of all women, and scoffed at Mary's attempts at furthering her education. Far from freeing her from the restrictions of her parents, marriage brought Mary nothing but further duties and discouragement.

Fortunately for us, Samuel Greig passed away young, leaving Mary enough money to comfortably live modestly on her own. She moved back in with her family and, though still having to raise her children and assume responsibility for the organisation of the household, she could now, for the first time in her life, learn at the pace she wanted. She was 26, and had the equivalent of a 15-year-old's education in mathematics – the basic principles of algebra, a solid foundation in geometry, but as of yet no trigonometry, function theory and certainly no calculus.

With the shackles off, however, Mary flew through all of the known fields of mathematics, and by the age of 30 had earned a silver medal for her solution to a problem in the *Mathematical Repository*. One year later, in 1812, she married her cousin, the adventuring diplomat William Somerville, who was as kind and supportive as the loathsome arch-fiend Samuel Greig was cold and imperious. For the rest of his life, he dedicated himself to helping Mary however he could – in running down rare maths texts at libraries, in copying her manuscripts, and in organising trips to introduce her to the scientific elite of Britain and the Continent.

Actively encouraged for the first time in her life, and having picked up French (again, self-taught), she waded into the heart of French mathematics which had, since the mid-eighteenth century, grown to dominance (Euler's titanic contributions notwithstanding) under the steady brilliance of Lagrange, Poisson, Fourier, and especially the reigning genius of Laplace.

Laplace's *Mecanique Celeste* was to the early nineteenth century what Newton's *Principia* was to the late seventeenth – a magisterial accounting of the motions of the solar system harnessing the most powerful mathematical tools available. Newton, realising that his audience could only be expected to trust and grasp so far the techniques of the calculus he invented, couched most of his arguments in pure geometric terms. Laplace, benefiting from the work in algebraic and functional analysis of Lagrange and Euler, was able to solve problems of greater difficulty and so to provide a breath-arresting, unified view of the long-term stability of the solar system.

Meanwhile, England had been dutifully spinning its wheels, completely out of sync with the dizzying speed of mathematical developments in France. Mary, however, had travelled to France and discussed Laplacean physics ... *with Laplace*. She was hailed throughout Europe for the profundity of her understanding in the deepest realms of mathematical physics, and when she at last returned to England, she was earnestly asked by Lord Brougham to prepare a work explaining Laplace's theories to an English audience.

She began in 1827, at the age of 47, and did not complete the work until 1831. The resulting book, *Mechanism of the Heavens*, was a masterpiece that not only presented a translation of Laplace's original two-volume thunderbolt but expanded it, filling in the sections where Laplace had somewhat condescendingly placed, 'it obviously follows that ...' when it was not obvious to anyone besides Laplace at all, and adding her own clear explanations of the consequences of Laplace's thought.

The book was a magnificent success, eventually selling an astonishing (for the time) 15,000 copies and securing Somerville's place in the first rank of British scientific minds. After a triumphal return tour through France, Mary settled down to write her second book, *On the Connexion of the Physical Sciences*, a tour de force review of all the cutting-edge work currently being done in the physical sciences, with Mary explaining the interconnection of all this astonishing new knowledge. It contained everything from the latest discoveries about the connections between electricity and magnetism to the gravitational consequences of the Earth's oblate spheroid shape. The book tore through multiple editions and served as an introduction for a new generation of British scientists into the emerging mysteries and puzzles of experimental and theoretical science. James Clerk Maxwell, the giant of late nineteenth-century physics, praised the book explicitly for its role in reinvigorating British scientific interest.

Somerville was already 56 by the time *Connexion* was published, an age when, statistically, she should have been either dead or at the very least far beyond any sort of creative prime. And yet, she continued to study and write up to her death at the age of 92. In her third book, *Physical Geography*, written in 1848, she risked the wrath of the established Church by advocating on behalf of the Old Earth geologists in the first ever English-language popular review of geology. Then, in 1869, at the age of EIGHTY ... NINE ... she published *On Molecular and Microscopic Sciences*, the least successful of her four major works. Whether she was too old to be in touch with modern developments, or whether she was simply too ambitious for the times (just try to describe molecular behaviour without using the words Electron, Proton, Nucleus, Polarity or Bond and you'll get a notion of the difficulties involved). She wasn't happy with it and it never caught the public imagination in the same way as her first three works.

As an original researcher, her work on the relationship between light and magnetism was accurate, with a nose for what the Next Big Thing was going to be, though ultimately her conclusions were shown to be flawed. As an epicentre of British, indeed European, scientific life, she reigned confidently for four decades. Faraday and Young, Laplace and Herschel, all respected her achievements and counted her as a warm and ceaselessly modest friend. She was honoured by the Royal Geographic Society at home, and the Italian Geographic Society abroad, given a pension from the British government for her contribution to English intellectual life, and a statue of her was commissioned by the British Royal Society. After four decades of constant struggle and intellectual deprivation, and five more of domestic happiness, international acclaim and blissful pursuit of the eternal truths of mathematics, Mary Somerville died in 1872.

FURTHER READING: In her nineties, Mary Somerville wrote her memoirs, which are a mixture of charming anecdotes and seemingly endless social gatherings of people only the most dedicated of early nineteenth-century European enthusiasts will have heard about. The additional notes compiled by her daughter also add a nice feeling for what Mary was like on a day-to-day basis which is equally lovely. Then, of course, she is also featured along with Émilie du Châtelet in Robyn Arianrhod's still terribly titled but thoroughly wonderful *Seduced by Logic*, which gives you two great mathematicians for the price of one.

Chapter 10

Sofia Kovalevskaya: Love Makes all the Partial Difference

Everybody needs love, but for some the striving after it so dominates their every action and decision that it becomes impossible to ever truly find it. Veering between professions, friendships and lovers, their desire for perfect love driving away by its intensity anybody who might have offered it, those possessed by such a need rarely live happily or end well, but their lives dazzle as against the more steadied demands of their contemporaries who settled for reasonable affection. In the history of mathematics, it would be difficult to find a person whose life and work was more impacted by devotion to the idea of an idealised, impossible love than Sofia Kovalevskaya (1850–1891), the nineteenth-century Russian mathematician who contributed equally to the theory of differential equations and the corpus of Russian literature, and whose cold end came too soon.

She was an anxious child, given to terrifying night visions and fits of panic in the face of deformity. A wax doll with a missing eye or the mere mention of a child with two heads would haunt her dreams. She was the second daughter in a household that had eagerly expected a first son, and lived her childhood in the shadow of that fact. While her parents doted on her older sister, and showed off her younger brother, Sofia they were content to leave be.

Armchair psychologists will see in this the germ of her lifelong profound quest for love – a latter life attempt to find that which was denied her as a child. Perhaps that is true, and perhaps her recollections of her childhood were distorted to take the shape of subsequent needs.

Something was lacking, though, and as a teenager she and her older sister were inspired by a Russian youth culture that held heroic self-development as an idealism-driven call to action. Sofia's father objected resolutely to her attending any university to study mathematics, so she did what so many other young Russian women were doing: she found a philosophical young man willing to marry her, bring her to a European university town, and then leave her be. He was Vladimir Kovalevsky, and his end would be as tragic as hers, though it would come much sooner.

She sneaked out of her parents' home to Vladimir's apartment, leaving behind a note of her intention to marry him. She knew that, merely by being alone in the same house with a young man for a few hours, the couple would *have* to get married by the rules of propriety. Bowing to her fait accompli, her father sanctioned the match and off they went to Europe, where Sofia began her career as a full-time student of mathematics.

She lived in cheap lodgings with a friend, her husband visiting her from time to time, as his own studies would allow, while she devoted herself fully to the study of her topic, reading day and night as she caught up with the most current trends in mathematical analysis. She eventually worked her way to the University of Berlin, where the great Karl Weierstrass was crafting those mathematical miracles that The Initiated still talk of with hushed awe. He saw the promise of her intellect and brought her into the department against the protest of the more conservative faculty.

She wrote three papers for her doctoral work, the most noted of which gave us the delightful Cauchy-Kovalevskaya Theorem. In it, she proved generally what French mathematician Augustin-Louis Cauchy (1789–1857) had only proven for a special case, namely that, if you have a partial differential equation involving an analytic function (that is to say, one which can be represented as a sum of powers of a variable) composed of x, t, and partial derivatives thereof restricted in degree by the original PDE, and if the initial conditions of the desired solution are themselves analytic functions of x, then a solution to the original differential equation exists in the neighbourhood of 0.

As a mathematics nerd, I have always loved the coyness of Existence Theorems – they hold out the guarantee of the existence of a solution without bothering to give you what you need to actually find said solution. On the strength of her papers, Kovalevskaya was easily awarded a doctorate, and so became the first woman in Europe since the Renaissance to hold such an advanced scientific degree.

The celebration, however, was short-lived. Summoned back to Russia by the death of her father, she gave up mathematics for some time in order to attempt a life of normality with her theoretical husband, Vladimir, to find in him at last the love that neither intimate friends nor mathematical study was able to provide. They had a daughter together, integrated into cultivated society, and Sofia caught the bug for financial speculation, which was well-nigh required of every late nineteenth-century Russian person of note.

She dragged Vladimir with her from scheme to scheme and, after nearly bankrupting the family, swore off anything smacking of fiscal adventure. Her husband, however, once hooked could never quite shake the urge to make one last big score. He latched on to a scoundrel who talked a good spiel about a wildly profitable venture while secretly filling Sofia's head with lies about Vladimir's sexual conquests in an attempt to split the couple.

It worked. Devastated, Sofia left her daughter in the care of friends and relatives and fled to Western Europe. Vladimir and The Scoundrel carried on until the latter's sudden but inevitable betrayal. Alone, bankrupt, and without the ability to take pleasure in the science that had soothed him in earlier troubled times, Vladimir took his life in 1883.

Sofia, meanwhile, was being courted by the University of Stockholm to become the first female professor of mathematics in Europe. She accepted with zeal at first, but as the years rolled on, she yearned increasingly for the more inspiring intellectual company of Paris, Moscow and Berlin, and viewed her lecturing duties in Stockholm as a sort of purgatory to be slogged through for the sake of money and prestige.

She sank into prolonged periods of absolute apathy, doing needlework for hours on end and reading novels to pass the time, while yearning for an all-consuming connection with another human being. Then, suddenly, a new and overwhelming love burst into her life. While her friends urged her to submit her new mathematical ideas for the prestigious *Prix Bordin*, she was occupied with her desperate love of a man whose great pleasure in life seemed to be the weekly breaking of her heart. She was torn between her desire to put her intellectual ideas down on paper and her guilt about not surrendering herself completely to love. Finally, though, begrudgingly, the work was done, and in 1888 she won the prize easily.

The work established a third type of integrable rigid motion, the case of a precessing top with moments of inertia (think mass, but instead of expressing resistance to linear motion, the moment of inertia is what expresses resistance to *rotational* motion) in a special ratio. The paper, capping the previous discoveries of Euler and Lagrange, was deemed so important that the award committee nearly doubled the monetary prize in recognition of its significance.

Alas, it was to be Kovalevskaya's last mathematical work. Her attention turned towards literature, and the more sparkling and sympathetic company offered by writers and journalists. Usually, the mathematician-turned-novelist is a recipe for disaster only matched by the actor-turned-musician, but Sofia's tragic, lonely youth, her burning sense of idealism, and the deep capacity for observation born by both, made her a writer instantly recognised as a voice of note for the coming generation. Her novels, *The Rajevsky Sisters* and *Vera Vorontzoff* (sometimes called *Nihilist Girl*), are full of her own unique perspective on vulnerability and purpose.

She was at work on other literary efforts, and was writing to friends about a new mathematics paper that would dwarf her previous work when, in 1891, an extended tramp through the snow while carrying her own luggage opened the door to a nasty case of influenza. Even while sick, she continued trying to fulfil her lecture obligations, but even her mercurial spirit couldn't overcome the disease, and she died on 10 February, alone in her room.

She had always lamented to her closest friends that nobody had ever truly loved her and yet, when the news of her death was announced, all the corners of the world flooded Stockholm with messages of condolence. Cartloads of flowers covered the coffin at her funeral, while a women's organisation in Russia raised a special fund to erect a monument to her memory. Today there are poems and novels, scholarships and lunar craters, dedicated to Sofia Kovalevskaya, the mathematician, the novelist, the teacher, who believed she had never known a single day of true and reciprocated love.

FURTHER READING: There are quite a number of books about Kovalevskaya now, though my favourite is probably still that written by her good friend A.C. Leffler back in 1894. Sofia had always believed that she would die young, and made Leffler promise to write her biography after her passing. It is called *Sonya Kovalevsky*, and is heavier on Sofia's literary production than her mathematical output, as Leffler resolutely understood the former and had not the slightest notion of the latter. For the mathematics part of her work, you can find bits and pieces of it spread throughout PDE texts. I have always been interested in Roger Cooke's *The Mathematics of Sonya Kovalevskaya* (1984), but as hardback copies of it *start* at $1,000 that's not happening anytime soon. Now, if anybody would like to buy me that book in order for me to review it, I would not to be too proud to accept.

Chapter 11

The Englishwoman in America: Charlotte Angas Scott and the Development of American Mathematics

In 1885, when you heard the term 'mathematical epicentre' one of the last places that would have sprung to mind was the United States. Sure, it had a handful of locally distinguished minds to its credit by that point – figures on the order of Thomas Godfrey (1704–1749), Nathaniel Bowditch (1773–1838), Theodore Strong (1790–1869), Benjamin Peirce (1809–1880), George William Hill (1838–1914) and Henry Fine (1858–1928), but a half dozen notables strung out over the course of a century hardly a robust intellectual tradition makes, and by the late nineteenth century, the situation was getting frankly embarrassing. Fortunately, that is when there came to the US an Englishwoman who brought with her not only connections to the wider world of European mathematics but a zest for teaching, gift for original research, and know-how for building up national scientific institutions that soon turned America from a mathematical backwater to a bustling community.

That woman, Charlotte Angas Scott (1858–1931) was, like the astronomer Maria Mitchell a generation before her and an ocean away, the beneficiary of her family's particular religious beliefs. Just as Mitchell's upbringing among Quakers brought her opportunities to educate herself and pursue a career in science not generally available to most young American women of her era, so was Scott's family's non-conformist Congregationalism the gateway for young Charlotte to whatever academic pursuits she found herself wishing to pursue. Her father, Reverend Caleb Scott, was a Congregationalist minister who was known for urging his parishioners to let their daughters 'not be all repressed and stifled in the iron mould of any conventionalism'.

Upon becoming the principal of Lancashire Independent College in 1865, Caleb had access to a bristling array of bright young minds whom he hired to provide Charlotte with tutoring in whatever fields excited her interest. On the strength of her mind and her private lessons, she won a Goldsmiths' Company scholarship in 1876 to attend Girton College, which had been established in 1869 under the name College for Women, at Benslow House (more popularly known as Hitchin College), but upon the college's move from Hitchin to Girton in 1872 to be nearer Cambridge, the institution changed its name. (Some sources say that Scott matriculated at Hitchin College, but by that point the name change had been in place for at least three years.)

Scott was one of eleven students entering that year, a positive explosion in attendance from the 1869 entering class of five individuals. At the time, Girton students were allowed to take classes at neighbouring Cambridge if granted permission by the class's professor, permission generally denied by about a third of the college's faculty, and often only extended under the proviso that the woman attendees would secrete themselves behind a screen so as not to be a distraction to the male students' delicate sensibilities. Girton students could also apply for permission to take the famous Tripos exam, though they could not be ranked among the male Tripos participants, nor receive a degree from Cambridge for their performance. Women before Scott had taken the Tripos, but Scott was the first to receive a score on the *fifty*-hour, *nine*-day examination that placed her among the top ten male performers. Her score entitled her to the title of Eighth Wrangler, but university rules prohibited her both from receiving that title, and from even attending the ceremony where the placements were announced and the titles awarded.

But then, something magical and rare happened. For those whose ideas about the maturity of British college men in the nineteenth century is based upon the stories of their appalling, self-serving, and brutish behaviour towards the Edinburgh Seven from 1869 to 1873 (for more on which, see Volume 1 of this series), the outcome of the 1880 Tripos ceremony will doubtless come as a profound surprise. As the Tripos placements were being read, everything proceeded as was traditional until the Eighth Wrangler position was called out, and the halls resounded with undergraduates yelling out 'Scott of Girton', drowning out the name of the man who had scored beneath her but was nonetheless receiving her title. Back at Girton, meanwhile, a special ceremony was arranged for Scott's accomplishment, wherein she was led between lines of Girton students, while they sang 'See the Conquering Hero Comes', and then seated on a dais, where they crowned her head with laurels and recited odes composed in honour of her victory.

Scott's feat made the English papers, and spurred the creation of a petition to let women sit the Tripos exam regularly, without the need to apply for special permission, a right which was granted in 1881. Though she had triumphed, Scott did not receive a degree for her work (Cambridge would not start awarding degrees to women until 1948), but thanks to the efforts of Sophia Jex-Blake, the leader of the aforementioned Edinburgh Seven, in 1876 England passed a law allowing (but not compelling) universities to award women degrees. Scott took advantage of the new law to receive her bachelor's degree with first-class honours from the University of London in 1882, and her PhD in 1885.

Meanwhile, Girton's hero student returned to her alma-mater as a lecturer in mathematics, serving in that capacity from 1880 to 1884, when she received an invitation from M. Carey Thomas to join the faculty of Bryn Mawr as an associate professor, one of the university's eight starting faculties, and one of only two women. There was a special sense of continuity for Scott at Bryn Mawr, as it was a college established by Quakers, who had a similar devotion to women's education to that of the Congregationalists of Scott's youth. With the founding of Bryn Mawr in 1885, they created the first American university offering graduate education for women, and Scott was happy to switch from the Girton-Cambridge atmosphere of intensive examination paranoia to the less student-mauling approach of America.

While at Bryn Mawr, she developed a reputation as an advocate for students who were applying themselves but faced various adversities, including an incident where she rose in the defence of a student who had done a great deal of fine work prior to catching tuberculosis, but whom the university was contemplating not awarding a degree on a technicality about uncompleted credits. While in America, Scott picked up golf and added it to her love of lawn tennis, activities which, while not strictly proper to conservative circles, she nonetheless thoroughly enjoyed, even as most other aspects of life in America left her feeling somewhat isolated and depressed, causing her to return to England at every opportunity to reconnect with colleagues and with her cultural roots.

Beginning in 1892, Scott published a series of papers on algebraic geometry that continued at the rate of one or two a year for the next decade and a half and which cemented her international reputation as a mathematician of note. She focused her work on plane curves of degree greater than two (i.e. equations in x and y featuring highest powers of three and greater), and in particular kept returning to the subject of singularities, which are technically defined as locations where curves behave very, very badly. These include cusps, points where graphs cross over each other, and isolated points that sit glumly by themselves away from the rest of the graph. Scott's work focused on characterising higher singularities, and in investigating the intersections of plane curves, through her own brilliant and individual sense of geometric objects and how to corral them within different structures that compel them to reveal their internal relationships.

In 1894, she published the textbook *An Introductory Account of Certain Modern Ideas and Methods in Plane Analytical Geometry*, in which, employing her clear writing style and gift for creating new ways to consider spaces and the objects inside of them, she was able to communicate some of her time's newest ideas about projective geometry, invariance and absolute conics in a manner that required virtually no revisions upon the text's re-publication three decades later. This was also the era when she was employing her experience as a member of some of Europe's most important mathematical associations to improve the cohesion of American mathematical life, joining the New York Mathematical Society (founded 1888) in 1891 and acting as one of the guiding forces behind its expansion into the American Mathematical Society in 1894, serving on its council for most of the 1890s, and acting as vice president in 1905–06.

When she first arrived at Bryn Mawr, it was on the understanding that, as the years went on, her teaching and grading load would be reduced so that she would have more time for research and administrative work. This did not ultimately happen, and Scott's frustration with the amount of time taken up by grading led her to explore the possibility of national standards for mathematical testing, taking up the role of Chief Examiner in Mathematics for the College Entrance Examination Board in 1902 and 1903. (Today the CEEB is known simply as The College Board, an institution now generally more reviled than not, but in Scott's day it was seen as an organisation that was helping to increase access to higher education for high school students.)

Scott was also central to improving the general mathematical expectations of the US college system. At a time when Harvard was not even particularly concerned whether its students entirely knew how to multiply, Scott was establishing at Bryn Mawr standards for general mathematical literacy to be met by all students, as well as expected competencies for mathematics majors, that are now commonplace in university curricula.

In 1906, Scott's publication career came to an abrupt end when she was diagnosed with rheumatoid arthritis, and advised by her doctors to spend more time engaging in outdoor exercise and gardening, and less time hunched over mathematical proofs. This, along with the serious advancement of a deafness that had begun during her Girton days, made teaching increasingly difficult for Scott, who by her retirement in 1924 was effectively entirely deaf. She returned to England in 1926, where she spent time happily among her relatives, and even returned to research with a paper on higher singularities that ranks among the most influential of any she wrote during her career.

Scott died on 10 November 1931 in Cambridge, and was buried in grave 4C52 beneath a headstone that does not mention her career as an internationally respected mathematician. That is unfortunate, but in truth she did not need a graveyard inscription to ensure the continuation of her memory, for her legacy lay elsewhere, in a Bryn Mawr mathematics department that grew on her watch to be a trailblazer in more rigorous undergraduate curricula, in her crystal-clear exposition of the lives and secrets of plane curves, and in the living testament to her example carried on by her seven PhD students, including the likes of future luminaries Ruth Gentry, Ada Maddison, and Marguerite Lehr. She helped bring a nation into the global mathematical community, and the memory and gratitude for that will last long past any mere etching in stone.

FURTHER READING: Charlotte Angas Scott is a figure you must find in pieces, here and there. For the content of her mathematical work, the best source by far is F.S. Macaulay's 1932 obituary piece in the Journal of the London Mathematical Society, an appreciation that the august institution will let you look at for forty-eight hours online if you pay $12, and will let you actually print out for yourself if you pay $49.

Forty ... nine ... dollars. For an obituary ... from 1932.

Absolutely scandalous, but if you're a big Scott fan, and pretty familiar with algebraic geometry, it is definitely worth a look. Her 1894 book, *An Introductory Account of Certain Modern Ideas and Methods in Plane Analytical Geometry* is available in a nice amply-sized reprint from the Leopold Classic Library, and it is a pretty fun trip that keeps challenging you to think of familiar objects in new ways. For her life and professional impact, both M.C. Bradbook's 1969 *That Infidel Place: A Short History of Girton College*, and the ever-indispensable *Women of Mathematics* edited by Grinstein and Campbell are useful, while you can glimpse more of Scott and the system she put in place in books about Bryn Mawr, and particularly about its influential and controversial second president, M. Carey Thomas, including Edith Finch's *Carey Thomas of Bryn Mawr* (1947) and Helen Horowitz's *The Power and Passion of M. Carey Thomas* (1999).

Chapter 12

Spherical Triangles and Domineering Males: The Saga of Grace Chisholm Young

When Grace Chisholm, at age 28, married the mathematician William Henry Young, she had every prospect of a brilliant career before her. The previous year she had been the first woman to earn a doctorate in Germany thanks to her influential work on the categorisation of spherical triangles, and a few years prior to that she had scored first place in the gruelling Oxford mathematical exam and earned first-class placement in the infamous Cambridge Tripos exam. A student under the wing of Felix Klein, one of history's greatest mathematicians, she could look forward to a lifetime of first-rank discoveries.

You may think you know this story after having heard it so many times over the course of the other books in this series: a brilliant woman on the verge of masterful greatness marries unwisely and suddenly abandons all intellectual activity on the say-so of a condescending, self-centred husband. It was the story of Isabel Morgan, of Harriet Brooks, of Mileva Maric Einstein, and of Mary Somerville before her first husband did the world a massive favour by dying young. And while there are elements of that tale in Chisholm's story, the web of expectations and neglect she found herself in upon marrying was a truly complicated tangle, giving her alternately complete creative freedom and complete subservience to the demands of her husband's mathematical visions and personal demons.

For as complicated as her adulthood was, the first stages of Chisholm's life were cheerfully stuffed with simple good fortune. Her father was Warden of the Standards and had been the head of Britain's Department of Weights and Measures, a man of influence and wealth who also believed fully in the talents of his daughter and in providing opportunities for their fostering. Her mother was a gifted pianist who regularly played duets with Chisholm's father, an amateur violinist. Chisholm's most treasured early memories were of exploring the British Mint with her father and working alongside him in his shop at home.

Happily ensconced in a creative, intellectual family, Chisholm did have an unhappy tendency towards terrifying nightmares and headaches, which had one overwhelmingly positive outcome: the family doctor recommended that, so as not to overtax her, she should only be educated in the things that she actually wanted to learn. Freed at the stroke of a doctor's pen from the smothering expectations that characterised most wealthy young ladies' early years, Chisholm set a course for herself that concentrated on her two great loves: mathematics and music.

Eventually, as the nightmares subsided, a governess was employed to fill out her education, and in spite of not having the benefit of a formal grammar school education, she managed to easily pass the Cambridge Senior Examination at age 17. She at first wanted to go on to study medicine but her mother sternly vetoed that plan (later Chisholm would characteristically study it anyway on the side, passing all the requisites except the internship), and so Chisholm returned to her long-standing companion, mathematics, thanks to a scholarship to Girton College, where she enrolled in 1889.

Girton, like Cambridge, relied heavily on a tutor system into which occasional formal classes were enmeshed, and Chisholm's tutor was a young man of fearsome reputation by the name of William Young. Gruff, sarcastic and demanding, he regularly reduced students to tears in the name of preparing them for the overwhelming Tripos exam. Their relationship at the time did not go beyond that of tutor and student, and Chisholm excelled in her programme of study, gaining a first-class placing in her 1893 exam.

It was a triumph, but it was also all England could offer her. There was no road to graduate studies available for women at the time, and so Chisholm faced either settling down to respectable charity work, as was widely expected of someone of her status, or finding an institution on the continent at which to study. Being Chisholm, she elected for the latter and set her sights on Göttingen, where the legendary Sofia Kovalevskaya had earned a degree *in absentia* in 1874. As it turned out, then, England's oppressive graduate system worked to Chisholm's advantage as it forced her out of England, where mathematics was caught in one of its recurring bouts of stagnation, and into Germany, where the makings of several mathematical revolutions were in the works.

One of the great names of Göttingen at the time was Felix Klein, whose work on the links between group theory and geometry was paving the way towards a new, modern, and more rigorously founded geometry. Importantly, he had no objection to taking on women students as long as they had already done work of value, and his influence was to be central in removing the bureaucratic obstacles to Chisholm's earning of a doctorate degree. Her work with him culminated in her dissertation, 'Algebraisch-gruppentheoretische Untersuchungen zur sphärischen Trigonometrie'.

Klein was so taken with her work that he included it in his classic text, *Elementary Mathematics from an Advanced Standpoint*. Chisholm investigated spherical triangles – objects formed by connecting any three non-diametrically opposed points lying on a sphere. Each of these has three sides, labelled say a, b and c, and three angles, labelled say α, β and γ. Taking sines and cosines of these six quantities gives twelve values that characterise that triangle, three of which are free to be whatever they want, and nine of which are dependent on the values chosen for those other three. Of all the combinations of sine and cosine values, the set of all those that describe an actual, possible spherical triangle is called M_3. Klein's question was how do we characterise this group M_3? What algebraic equations are satisfied by the points in it? How can we represent any surface passing through it as a series of equations? How 'big' is it?

These questions struck at the heart of modern mathematics' attempt to marry group theory to geometry, and Chisholm's contribution was to think not in terms of sines and cosines of a, b, c, α, β, and γ, but rather in terms of the cotangents of $a/2$, $b/2$, and so on. This representation allows M_3 to be represented in a six- instead of twelve-dimensional space, and thereby led to the answer that M_3 is an order eight space completely representable as the intersection of three surfaces of degree two in six-dimensional space.

Chisholm's approach to spherical triangles represented a neat and original shift in the framing of the problem that provided basic answers to an emerging set of questions. Chisholm sent her work to her former tutor, who soon proposed marriage. She refused him at first, stating that she was afraid of what marriage might do to her nascent mathematical career, but Young, heedless – as usual – of anybody's concerns but his own, kept pressing his case and in 1896 they married.

And now the story gets complicated. Young had tremendous difficulty his whole life in securing a permanent post that would allow his family to settle, resulting in a yo-yo existence in which Young would find temporary work, leave home for an undefined amount of time, and then return again for another undefined amount of time before repeating the cycle. A brilliant man, he lacked focus and the slightest awareness that other people's problems might possibly be more important than his own, and when he was home he demanded every drop of Chisholm's time to organise his life for him and put his thoughts in order. He was so domineering in his usurpation of her attention that, whenever he left to assume a new post, Chisholm would habitually lie in bed for days, paralysed by headaches as her body sought to reset itself from the juggernaut of his presence.

When he was gone, however, she could carve out some time for herself in between the raising of six children, each of whom was to obtain some degree of fame as a mathematician or scientist (well, except the one who died in the First World War, but five out of six is still *pretty* good you'll admit). Chisholm still had to manage the demands for her mathematical insight that came from Young by post, including a self-serving decree that, for a time at least, their joint work should be published under his name only, as he had a greater need of the laurels than she. But letters can be folded and stuffed in a drawer in a manner that humans typically cannot, and so during those absences she managed to learn six languages, undertake medical studies, and, most importantly, work on her own mathematical problems, including her part in the Denjoy-Young-Saks Theorem.

DYS is a tremendously useful theorem which investigates that central question to all of calculus – at what points can the slope (or steepness) of a given graph be calculated? This is important, because by knowing the slope of a graph at a point, you know what the rate of change is there for the quantity depicted by the graph – How fast are these icecaps melting? How quickly is this asteroid speeding up? How precipitously is the bee population declining? These are all questions about rates and thus about slopes. The DYS theorem relies on objects called Dini derivatives which are great fun but as I have probably already exhausted everybody's patience with that spherical triangle bit earlier, I'll just say that they describe the extremes of what is happening to the slopes of a curve as you approach a point from the left or right. In 1915, Denjoy proved that, for any continuous function, there are only three possible ways for these derivatives to behave, only one of which allows the slope to be found. So, if you have eliminated the two options that don't allow the slope to be found, then it must be the case that the slope is calculable. In 1917, Young extended this result to a broader class of functions, so-called 'measurable' functions, which can be very weird and theoretically tricky to deal with, but which, looked at through the Dini derivatives, can be analysed much more simply. And that is pretty great.

All told, between her solo papers and her joint papers with Young, she published some 220 articles over four decades, in addition to the *First Book of Geometry* (1905) which aimed, through foldable cut-out figures, to reacquaint children with the physically graspable truths of solid geometry, and a book called *Bimbo* (the nickname of her baby at the time – bimbo being a simple diminutive of 'bambino' which didn't take on its modern connotations until the 1920s), which described cell division via a children's short story.

By 1940, Chisholm had spent forty years in continental Europe at the whim of Young's shifting academic fortunes, but the rise of Hitler was a clear indication that it was time to return to England. Chisholm ventured back in 1940, but Young vacillated, afraid that he would be judged for his many years of close work with German universities. As it happened, he waited too long, and the window of opportunity for return closed. Young died in occupied Europe in 1942, senile and unbalanced in his isolation. Chisholm died two years later, at the age of 76, just before Girton College was to unveil its plan to award her an honorary fellowship in recognition of her life's work.

Grace Chisholm Young's life is equal parts inspiration and cautionary tale. Her solo achievements speak for themselves, but there is a great lurking What If to be tackled. What If Grace Chisholm had resisted William Young? Certainly, his career would have suffered – by his daughter's accounts he was utterly dependent on Chisholm to sift, focus and organise his ideas – with the result that mathematics would lose many of his published works and all their jointly published ones. But in compensation we would have been treated to forty years of a uniquely multifaceted mind entering whatever fields might catch her fancy and I cannot help but think that, in the summing, that would have worked towards a net positive. As it is, we have several remarkable individual results, hundreds of joint works, two thoroughly individual books, and a small gaggle of offspring each of whom (well, five out of six of whom) advanced scientific knowledge in their own right, and while that might not represent all that ought to have been, it is still a staggering testament to a mind capable of bringing mathematical order from the chaos of reality.

FURTHER READING: The main biographical source for Grace Chisholm Young is Grattan-Guinness's eighty page 'A Mathematical Union: William Henry and Grace Chisholm Young' in *Annals of Science* 29 (1972). But those are hardly lying around anyone's local bookshop, so for shorter accounts that have quite different focuses you can pick up either the omni-useful *Women in Mathematics: A Biobibliographic Sourcebook* (1987) by Grinstein & Campbell, or Teri Perl's unique *Math Equals: Biographies of Women Mathematicians* (1978). Klein's book is available in an easy-to-find Dover reprint and is thoroughly wonderful.

Chapter 13

Brief Portraits: The Eighteenth and Nineteenth Centuries

Diamante Faini (née Medaglia) (1724–1770)

Born in Brescia on 28 August 1724, Diamante Medaglia studied theology, Latin and religious history under the tutelage of her uncle, a Catholic pastor, while devouring privately the poetry of the sixteenth- and seventeenth-century poets who sought to reclaim the heights scaled by fourteenth-century poet and philosopher Francesco Petrarca. The brilliance of her love sonnets gained her admission to several prestigious academies and literary societies, but her father, a doctor, took issue with the composition of love poetry as a life direction for his daughter, and married her off to Pietro Faini, a physician acquaintance of his.

After marrying, she forswore the production of any more sentimental poetry, and instead trained the intensity of her intellect on the study of science and mathematics. She studied Euclid's *Elements* under the instruction of Giovanni Suardi, and in 1763 stood before the University of Brescia to make a spirited plea for the inclusion of classical philosophy, mathematics and science in the education of all young women.

Maria Angela Ardinghelli (1730–1825)

Ardinghelli is known to us today primarily as the mathematician and physicist who translated the works of Stephen Hales (1677–1761) into Italian. Hales was a Newtonian physiologist of the early to mid-eighteenth century, most famous as the first European scientist to experimentally measure blood pressure, but also as an experimenter whose calculations and experiments on plant transpiration, root pressure, and blood volume pumped by the heart stood as important, if often gruesome, examples of how to bring mathematical rigour to the biological sciences. Ardinghelli discovered and corrected the mathematical mistakes in both the English

50

original, and French translations then available, and provided explanatory footnotes that laid out the calculations underlying Hales's work for the reader. Born in Naples, where she remained the rest of her life as the hostess of one of the city's most sought-after intellectual conversation societies, and correspondent with the Paris Academy of Sciences. She is known to have performed experiments in the manner of Hales to further the use of mathematics in explaining biological and physical phenomena.

Wang Zhenyi/Chen-i (1768–1797)

Wang Zhenyi had the warm fortune of being born into a family of distinct brilliance and intellectual openness. Her grandfather, Wang Zhefu, was a former governor and the owner of a seventy-five-shelf library of learned volumes. Her father, Wang Xichen, was a doctor and medical author, and her grandmother was a student of classical poetry. Each contributed the best of their wisdom and experience to teaching young Wang Zhenyi about the worlds of astronomy, poetry, medicine and mathematics. Upon the death of her grandfather, she spent five years in Jiling, where Zhenyi studied archery, equestrianism and martial arts with the wife of a Mongolian general.

Around the age of 18, Zhenyi concentrated her studies around astronomy and mathematics, achieving fame in the former field for her writings about equinoxes and lunar eclipses, and in the latter for her published works on trigonometry and calculation, including the book *The Simple Principles of Calculations,* and the article 'An Explanation of the Pythagorean Theorem and Trigonometry'. She also undertook the task of rewriting late seventeenth-century mathematician Mei Wending's *Principles of Calculation*, simplifying the explanations to make the material more widely comprehensible to beginning students. She died of unknown causes at the age of 29, her manuscripts preserved and published by her nephew Qian Yiji. In 1994, a Venusian crater was named in her honour. For a fuller account of her life, see the second volume in this series, *A History of Women in Astronomy and Space Exploration.*

Elizaveta Fedorovna Litvinova (née Ivashkina) (1845–1919?)

An almost exact contemporary of the more famous Sofia Kovalevskaya, Elizaveta Litvinova had a life perhaps equally tragic. Elizaveta Ivashkina was born in 1845 in Tula, Russia, to Fedor Ivashkin, a member of the landed gentry who was at least progressive enough to send his daughter to St Petersburg to study at a girls' high school. The next logical step for a young woman of ability was to study at a university in Europe, as Kovalevskaya had done. Unfortunately for Elizaveta, the man she married in 1866, Viktor Litvinov, was completely against the idea of his wife receiving a European education, and refused his permission for her to leave the country to do so. Under the laws of Russia at the time, that decision sealed her fate, as women were unable to obtain passports out of the country without permission from their fathers or spouses. So, she did the only thing she could do.

She waited for the bastard to die, while taking private lessons in mathematics from Aleksandr Strannoliubski. Viktor helpfully passed away in 1872 and that year Litvinova moved to Zurich to study at the Eidgenossische Technische Hochschule, a polytechnic institution, instead of the university there. She was finally ready to really begin her academic career when, in 1873, Tsar Aleksander II decreed that all Russian women studying abroad must return home, or face the end of any possible academic career inside of Russia. Litvinova defied the tsar's order and remained in Zurich, completing her undergraduate degree by 1876, and subsequently attaining her doctorate in Bern in 1878 for her dissertation on a problem involving the mapping of a curve on to a simple circular surface.

Returning to Russia in 1878, she found that the authorities were keeping to the letter of the tsar's 1873 decree, and for nine years the only teaching post Litvinova could find was as a maths teacher in the lower classes of a girls' academy. It was low-paid, by-the-hour work, and to supplement that income, Litvinova began writing biographies of popular mathematicians and over seventy articles on mathematical pedagogy. Litvinova retired in 1917, and probably perished during the famines of 1919 caused by the Russian Civil War.

Christine Ladd-Franklin (1847–1930)

Known primarily for her contributions to the theory of colour vision today, in her time Christine Ladd-Franklin was also known for her skills as a logician. The daughter of a New York merchant, who lost her mother at the age of 12, she spent her adolescence with her paternal grandmother, graduating her high school in 1865 as class valedictorian. She received her bachelor's degree from Vassar College in 1869, during which time she studied with the astronomer Maria Mitchell, but decided ultimately on a degree in mathematics, as something she could pursue more easily independently, given the restrictions on women in physics and astronomy of the time.

From 1869 to 1878, she taught at a secondary school and wrote seventy-seven mathematical problems for the *Educational Times*, and nine articles for mathematical journals. In 1878, she applied for admission to Johns Hopkins University as a graduate student, though the college was not at the time open to women students. Her application, however, was seen by James Sylvester, who was familiar with her publications, and he asked the administration for special dispensation allowing Ladd to attend only his classes, a permission that was eventually extended to attending the lectures of Sanders Pierce and William Story. By 1882, she had completed all of the work necessary to receive a doctorate, but was not awarded one by Johns Hopkins until 1926, though she was allowed to lecture there in logic and psychology from 1904 to 1909.

Beginning in the mid-1880s, Ladd (now Ladd-Franklin since her 1882 marriage to mathematician Fabian Franklin) began her studies of colour theory, and the stages of colour recognition in the human eye and brain (about which more in the *Psychology* volume of this series), while also producing influential papers on logic, including 'The Algebra of Logic' (1883) and 'On Some Characteristics of Symbolic Logic' (1889).

Ellen Amanda Hayes (1851–1930)

Ellen Hayes was the sort of academic spoken of in hushed, cautionary tones by her colleagues; a completely individualistic presence dedicated to radical causes without any thought for the consequences to her reputation or career. It was a nature that she developed early, when her Ohio family let her run more or less feral on her grandparents' farm, climbing trees, skating, and in short doing all the things that a proper young lady ought not to. Her mother, a teacher, encouraged her interests in plants and astronomy, while the best primary education available to her was a simple one-room, ungraded school in the best (and worst) rural American tradition.

Nevertheless, in 1875 she was accepted to attend Oberlin College, receiving her bachelor's degree after three years, whereupon she started her career as a mathematics professor at Wellesley College. She quickly earned a reputation as a radical and freethinker, attending church only rarely, wearing practical clothes instead of the elaborate long skirts of the era, supporting labour unions and temperance movements, marching in protest against the 1927 execution of Sacco and Vanzetti, and raising money for Russian orphans in the middle of the United States' first Communist Scare. Her outspokenness probably kept her from earning emerita status upon retirement, though in 1897 she was made chair of the newly developed Applied Mathematics department at Wellesley. She published four mathematical textbooks during her career, and wrote on issues of mathematical pedagogy in addition to her articles on suffrage and social reform. In 1929, at the age of 78, she began a new career teaching at a school for women industrial workers, before passing away in 1930.

Alicia Stott (née Boole) (1860–1940)

The first third of Alicia Boole's life was not a happy one. She was born the daughter of a mathematics professor at Queen's College in Cork, Ireland who died when she was 4 years old. After her father's death, she was selected from among her four siblings to live with her uncle, where for seven years she lived an unspeakably lonely and drab life, before moving back in with her mother in London at the age of 11, where her condition was scarcely improved, crammed together with an invalid mother and four siblings in a filthy one-room apartment without access to light or mental stimulation.

At 16, Boole was packed off to Ireland to earn some money as a probationer at a Cork children's hospital, only to return to London again, where in 1890 she married Walter Stott, an actuary, and had her interest in mathematics stoked by a friend of the family, Howard Hinton. Hinton introduced Stott to the subject of polytopes, which you can think of as multidimensional generalisations of our friends the three-dimensional polyhedrons. Stott was highly regarded for her ability to imagine and construct four-dimensional polytopes, and for the development of Expansion and Contraction operators, which allow mathematicians to generate interesting high-dimensional polytopes from more basic examples. She published her results in two influential papers, 'On Certain Sections of the Regular Four-Dimensional Hypersolids' (1900) and 'Geometrical Deduction of Semiregular from Regular Polytopes and Space Fillings' (1910). For two decades, she relinquished her public mathematical researches to devote her time to the role of housewife, until in 1930 H.S.M. Coxeter convinced her to re-enter the mathematical fray, whereupon she discovered two constructions for a four-dimensional, 96-vertex, object she dubbed the 'snub 24-cell'. She and Coxeter continued working together until her death in 1940.

Winifred Haring Merrill (née Edgerton) (1862–1951)

Winifred Edgerton was the first American woman to receive a PhD in mathematics, and the first woman to earn a doctorate of any sort from Columbia University. The precise details of her early life are somewhat hazy and contradictory, but from them we can discern that she was taught by private tutors, provided with her own at-home observatory to nourish her gifts for astronomy, and was from a family that counted such American luminaries as James Russell Lowell and Oliver Wendell Holmes as friends.

In 1883, Edgerton earned her bachelor's degree from Wellesley and independently calculated the orbit of the Pons-Brooks comet from the Harvard Observatory's data. Approaching Columbia University, she asked for permission to use its telescope and attend certain classes to further her studies, which she was allowed to do, provided that she cleaned the astronomical equipment and kept out of the way of the male students. She completed all of the work for a PhD in mathematics, but was going to be denied the actual doctorate degree – until she went to each individual trustee and pleaded her case. As a result, in 1886 she became the first American woman to earn a doctorate degree in mathematics, due to her dissertation on the analytic and geometric interpretation of multiple integrals, which also attempted to unify a number of different systems of describing mathematical coordinates.

Having achieved headlines with her PhD, Edgerton was offered a professorship at Wellesley, but had to turn it down, as in 1887 she married Frederick Merrill, and the American educational system of the era was resolutely set against the idea of women continuing in their academic professions after marriage. From 1906 to 1928, Edgerton directed the Oaksmere School for Girls, opening a Paris branch of the school in 1912. From 1948 to her death in 1951, she worked as a librarian in New York City.

Margaret 'Maud' Theodora Meyer (1862–1924)

Meyer was one of the first women to benefit from the doors opened by Charlotte Angas Scott, earning a score equivalent to that of the Fifteenth Wrangler in the 1882 Tripos examination, capping her three years at Girton College. After teaching at Notting Hill High School from 1882 to 1888, she began what was to be a thirty-year career as a professor at Girton College, where she wrote, but did not publish, a number of works in mathematical astronomy, and performed calculations for the Air Department in the closing year of the First World War. She was inducted into the Royal Astronomical Society the first year they opened membership to women, in 1916.

Helen Abbot Merrill (1864–1949)

Born 30 March 1864, in New Jersey, Helen Abbot Merrill's story is tied intimately to that of Wellesley College. She entered Wellesley in 1882, twelve years after the college's founding, with the initial intention of becoming a Classical language student. After her first year, however, she switched to the field of mathematics, graduating in 1886. For several years thereafter, she held different teaching positions until Helen Shafer, who had become president of Wellesley in 1888, asked Merrill to join the faculty in 1893. It was a low-paid position, and Merrill was kept at the associate professor level for thirteen long years, but it was better than the life of a roving high school teacher. She took some time off in the dawning years of the twentieth century to study at the University of Chicago, University of Göttingen, and ultimately at Yale, where she received her doctorate in 1903 for her dissertation, 'On Solutions of Differential Equations which Possess an Oscillation Theorem'. Returning to Wellesley, she became chair of the mathematics department in 1916. She wrote two textbooks with Clara E. Smith, *Selected Topics in College Algebra* (1914), and *A First Course in Higher Algebra* (1917), while carrying on her own research in function theory.

Clara Latimer Bacon (1866–1948)

The daughter of an American pioneer family, Clara Bacon was born on 23 August 1866, in Hillsgrove, Illinois. She graduated from Hedding College in 1886, and then received a second bachelor's degree, from Wellesley, in 1890. She served as a secondary school maths and science teacher until 1897, when she was invited to begin teaching at the Women's College in Baltimore, which had been founded in 1885, and which would become Goucher College in 1910. While teaching at the Women's College, Bacon worked on her master's degree at the University of Chicago, which she earned in 1904 for an investigation of the chords formed between two conic sections. Her graduate work was continued at Johns Hopkins, where in 1911 she became the first woman to receive a mathematics PhD from the university, for her work on Cartesian ovals (objects you get by declaring two foci, say A and B, and two constants, say m and n. The Cartesian oval formed by those four quantities is the set of all points where the distance from A to the point, plus m times the distance of B to the point, equals the constant n).

Meanwhile, at the Women's College, she became an associate professor in 1905, and then full professor in 1914, continuing in that role until her retirement in 1936.

Frances Hardcastle (1866–1941)

A native of Essex, Frances Hardcastle came from distinguished maternal scientific stock. Her mother was Maria Sophia Herschel, who was the daughter of astronomer John Herschel, granddaughter of the even more famous astronomer William Herschel, and great-niece of astronomy legend Caroline Herschel. Like many of the figures in our survey thus far, she received a Certificate in Mathematics from Girton College, passing the Tripos II in 1892, whereupon she went to the University of Chicago and Bryn Mawr (where she studied with none other than Charlotte Scott), before returning to Cambridge for her postgraduate work. Her main field of study was point-group theory, which focuses on groups of symmetries whereby at least one point remains fixed, writing 'Observations on the Modern Theory of Point-Groups' in 1897, 'A Theorem Concerning the Special Systems of Point-Groups on a Particular Type of Base-Curve' in 1898, and 'Present State of the Theory of Point-Groups' in 1900. She received her bachelor's degree from the University of London in 1903, and her master's from Trinity College Dublin in 1905. She was a supporter of the International Congress of Women, and advocate of women's suffrage.

Philippa Garrett Fawcett (1868–1948)

Another member of Women in STEM royalty, Philippa Garrett Fawcett was the niece of medical pioneer Elizabeth Garrett Anderson, and the daughter of Mildred Garrett, co-founder of Newnham College and a council member from 1881–1909. Her father was Henry Fawcett, the famous blind reformist MP from Salisbury who worked as Postmaster General in one of Gladstone's governments. From this rich intellectual soil Fawcett emerged, in 1890, to become the first woman ever to earn the top score on the Tripos exam, earning a score 13 per cent higher than the nearest male. Though still unable to earn the coveted title of Top Wrangler due to her gender, the feat made headlines, and earned Fawcett a scholarship that allowed her to carry out fluid dynamics research resulting in papers such as 'Note on the Motion of Solids in a Liquid' (1893) and 'The Electric Strength of Mixtures of Nitrogen and Hydrogen' (1894).

For ten years, Fawcett taught at Newnham until she was offered a chance to rebuild the mathematical educational system in the Transvaal in 1902. The region had just emerged from the various tragedies of the Boer War, and it was Fawcett's task as a lecturer at the Normal School in Johannesburg and an assistant in the Transvaal Education Department to train new maths teachers and erect new schools.

Returning to London in 1905, she earned a degree from the University of Dublin, and took up a position as chief assistant director of education on the London Central Council. From 1918 to her death in 1948, she was a fellow of University College. One month before her death, women were finally awarded the ability to earn degrees from Cambridge.

Ada Maddison (1869–1950)

Ada Maddison was born in Cumberland, England to Mary and John Maddison, and of her early life we know virtually nothing. Our first definite scrap of information comes from when she began attending the University of South Wales in 1885, before receiving a scholarship to attend, you guessed it, Girton College, beginning in 1889, entering in the same year as Grace Chisholm Young. On the end of year Tripos exam, she received a score equivalent to the twenty-seventh Wrangler, and went on to earn another scholarship, to study for a year at, you guessed it again, Bryn Mawr under Charlotte Angas Scott, before receiving her bachelor's degree from the University of London in 1893.

She then won a Mary E. Garrett Fellowship that allowed her to attend lectures at the University of Göttingen, which at the time featured the mathematical dream team of Felix Klein and David Hilbert, the former of whose paper, 'The Arithmetizing of Mathematics', Maddison translated and published in 1896. It was during this era that she published 'On Certain Factors of c- and p- Discriminants and Their Relation to Fixed Points on the Family of Curves' (1893) and 'On Singular Solutions of Differential Equations of the First Order in Two Variables and the Geometric Properties of Certain Invariants and Covariants of Their Complete Primitives' (1896).

Following her time in Göttingen, Maddison received her doctorate from Bryn Mawr in 1896, and took up a position as assistant secretary to Bryn Mawr's president, M. Carey Thomas. She remained at Bryn Mawr until her retirement in 1926, serving in a nearly exclusively administrative role by 1910.

Louise Duffield Cummings (1870–1947)

Pure mathematician Louise Cummings was born in Ontario, and received her first two degrees in Canada – a bachelor's in 1895 and master's in 1902, both from the University of Toronto – before migrating to that mecca of North American women's mathematics, Bryn Mawr, in 1898. Most of her career, however, was spent at Vassar, where she joined the faculty in 1902 as an instructor, ultimately becoming an assistant professor by 1915, an associate professor by 1919, a full professor by 1927, and a professor emerita in 1935. Her work in the early 1910s focused on triple systems, i.e. vector spaces over a field possessing a trilinear map, which have important connections to Lie and Jordan algebras. In the 1930s, she worked on hexagonal and heptagonal systems arising from seven and eight lines in a plane, respectively, and a system for comparing differently constructed straight-line nets.

Mabel Minerva Young (1872–1963)

Mabel Young began life the daughter of a music-loving mother, and a Massachusetts state legislator father with American ancestors traceable back to the early seventeenth century. Growing up in Worcester, Massachusetts, she attended an excellent high school that prepared her for attending Wellesley in 1894. Receiving her bachelor's from Wellesley, she went on to study at Columbia University, where she earned her master's degree in 1899, before returning to Wellesley as an assistant professor in 1904, continuing to teach at the college until her death in 1963, taking a break in 1914 to earn her doctorate from Johns Hopkins University.

Young's mathematical work focused on projective properties of curves and surfaces, including her 1933 study of the infinity of triangles that results when you take two fixed tangent lines to a parabola, and then look at the series of triangles formed by a variable tangent to the parabola that is free to move along it, and intersect with the original two fixed lines. The properties of the triangles so formed, and the important points associated with them, were the topics of her paper. Earlier, in 1916, she had written an article on Dupin cyclides, which are essentially inverted toruses or cylinders first discovered by Charles Dupin in 1803. She also solved five American Mathematical Society open geometric problems during her career, using analytic and projective geometry.

Chapter 14

Emmy Noether Solves the Universe

'Momentum is always conserved, except when it isn't.'

In secondary school physics, we learn all manner of conservation laws, one at a time, when they accidentally happen to pop up, without so much as a word of explanation for WHY nature seems to care so much about these quantities. We've asked, of course, only to have our knuckles rapped for impertinence or, in our less corporal age, we've been referred to Google to figure it out as best we can for ourselves.

One hundred and five years ago, a woman who was only begrudgingly allowed a university education gave us that very WHY and with it one of the most powerful tools in mathematical physics. Her name was Emmy Noether and she was born in Erlangen, Germany in 1882. Her father was a mathematician (one of whose important theorems was actually proven by Charlotte Angas Scott in 1899) and she too had a marked predilection for mathematics that only grew stronger as she delved further into its open mysteries.

In nineteenth-century Germany, a woman could only attend classes at a university with the express permission of each teacher. Every course that she wanted to take, she had to set aside time with the instructor and plead her case for being allowed to sit in the same classroom with the men, promising not to be a distraction and silently swallowing their regular advice to turn to more womanly subjects (Max Planck famously rejected all women applicants to his lectures out of hand ... until he met Lise Meitner).

Noether ran the gauntlet, however, with a steadfastness in the face of rank unfairness that would mark her entire career. She received her bachelor's degree equivalent in 1903 and wrote her doctoral dissertation (on bilinear invariant theory) in 1907 at the University of Göttingen.

As we have seen in previous portraits, this was *the* place to be for mathematics. David Hilbert was there. Hermann Minkowski was there. Felix Klein was there. Titanic minds who remain popularly unknown today because they did their work in mathematics rather than the sexier fields of physics or chemistry, they would also be Noether's friends and champions in her battle for recognition from the university.

Noether was not only in the right place but also studying the right field for her moment in history. She was an expert in invariant theory and group transformations, which govern how quantities change when you transform the coordinate system where they live. Newton had some assumptions about how such coordinate shifts altered measured values – assumptions which were blown apart in 1905 with Einstein's Theory of Special Relativity. In the fallout of that titanic event, mathematicians and physicists were looking for something that would link classical Newtonian conceptions of conservation with the new and strange world of relativity, and eventually with the even stranger world of quantum physics. Without such a unifying theory of conservation, physics threatened to fly apart into a chaos of special cases.

In 1915, Emmy Noether produced just such a theory, and published it in 1918 (a bonafide mathematics nerd, when asked about 1918, will get rather excited and start talking about Noether's theorem and then, perhaps, as an afterthought, recall something about the First World War ending that year too). And now, with your tender indulgence, I want to put on my maths teacher hat for a bit and talk about that very theory, because it is truly lovely and powerful, and once you wrap your head around it, the universe just shines with snazziness.

Noether's Theorem invokes a bit of specialised vocabulary. In particular, it tells us what quantities are preserved (momentum, energy, charge, etc.) for a particular physical situation whose coordinate system undergoes a particular transformation. So, for example, if you have a falling rock, and you spin the x and y axes 90 degrees around the z axis, what measured quantities come out just the same as when you measured them in the original, unspun system? Emmy Noether's answer encompasses every conservation law that went before, and anticipated all of those discovered since, even those in areas of physics she couldn't have begun to imagine from the vantage point of 1915.

Consider two points in space, or two events in space-time. There are lots of ways for an object to move from one to the other, but only one that minimises the difference between the Kinetic and Potential Energies (called the Lagrangian) for a particle making the trip. If you take that path, your KE and PE will be as balanced as possible, and we call that path 'extremal'.

Now, if you're on that path, and we shift the coordinate system around you (say, by rotating the x and y axes under you a bit), and the overall difference between KE and PE doesn't change, or changes only very slightly, then we say the motion is 'invariant' under that coordinate transformation.

If you tell me about an object undergoing a given motion, and how you want to change the coordinate system, Noether's Theorem will tell us exactly what conservation law *must* hold in that situation. What is phenomenal is that using this method, you can not only derive all of the conservation laws we're used to from secondary school physics, but a bunch of other things that you could not know from the older Euler-Lagrange equations and Hamilton Principle techniques that Noether fused in her own theorem. They explain d'Alembert's insights from a century and a half before just as easily as Feynman's ideas from three decades after her death. It was one of those grand moments in intellectual history when a shifting mass of unfathomable complexity solidified in three slick lines of text into a single over-arching theory about invariance and conservation and their role in shaping the development of the universe.

After publishing her groundbreaking work, the only material improvement Noether saw was that the university, under pressure from Einstein, Hilbert and Klein, finally allowed her to lecture to students under her own name (until then, her classes had to be done in Hilbert's name, as women were not allowed to lead classes). Of course, she still wasn't paid or officially recognised as a professor. In 1922, the best she got was 'unofficial associate professor' status and a small stipend for teaching abstract algebra, a field that she was making regular and foundational contributions to since publishing her theory.

Her life continued in this fashion, recognised by the greatest minds in physics and mathematics for her piercing insights into the theory of Lie groups and noncommutative algebras (the significance of which we are only just starting to unwrap now), but without an official position proportionate to her skills or renown. And so she marched on for eleven years, writing the laws that every abstract algebra student knows by heart, until 1933 when the Nazis came to power and she, of Jewish origin, was forced from her position. Unlike Lise Meitner, who fought to retain her place in spite of unceasing harassment at the hand of Nazi officials, Noether saw the direction of the wind and fled the country for a position at Bryn Mawr College, one that she occupied for less than two years before a failed surgery to remove an ovarian cyst ended her life at the age of 57 in 1935.

FURTHER READING: The beauty of Noether's Theorem is almost impossible to fully appreciate until you see it at work, churning through problems in wildly separate fields of physics with the same elegant ease. My own appreciation of the Theorem's far-reaching applicability was fostered by Dwight Neuenschwander's delightful book, *Emmy Noether's Wonderful Theorem*. His build-up to the theorem itself requires really only a first-year college calculus level of maths fluency – if you're cool with the chain rule for partial derivatives, you're probably ready. After that, things get turned up a notch as he applies the Theorem to different fields, but he is very good at walking you through the thinking, and his insights into the historical development of invariance theory and the calculus of variations are clear and invaluable.

Chapter 15

Hilda Geiringer and the Curious Behaviour of Stressed Metals

Beholding a bar of metal, it seems an object almost primal in its simplicity. Solid, reliable, the stuff of which cities are made. Peek beneath the surface, however, and what you see is a miniature universe of swirling complexities, of tightly packed atoms and delocalised electrons clambering hither and thither with no set home, behaving in ways that can only be described statistically. That resting slab of matter is a mess mathematically, and when you put it under stress, a whole new range of bedevilments open up.

Fortunately, in the early twentieth century, a mathematician happened along who was able to significantly tame the chaos of stressed metals, publishing the equations for metal deformation under stress that still bear her name only two years before Nazi persecution compelled her to flee her homeland. Hilda Geiringer (1893–1973) was born in Vienna on 28 September 1893, to a middle-class Viennese Jewish family. Her mother, Martha Wertheimer, was Austrian, and her father, Ludwig, a textile manufacturer, was either Hungarian or Slovakian, and both of them believed powerfully in women's education, sending Hilda to the University of Vienna to study under Wilhelm Wirtinger after she showed prodigious mathematical talent in high school.

Wirtinger was a founding figure in spectral theory with broad interests in mathematics who had a knack for attracting some of the century's greatest minds as his students, including Erwin Schrödinger, Kurt Gödel and Olga Taussky-Todd. Geiringer studied at the University of Vienna from 1913 to 1917, earning her doctorate in 1918 on the strength of her dissertation, 'Trigonometrische Doppelreihen', which dealt with double trigonometric series. In 1807, Jean-Baptiste Joseph Fourier had discovered how to represent any smooth function as a series of sines and cosines, which remains one of our most beloved of mathematical tools. Any time we can take an ill-behaved function and build it out of a set of nice, well-understood, well-behaving functions like sines and cosines, we tend to be very happy in mathematics, and Fourier's series (which we now call Fourier

Series) helped him in 1822 to solve the long-standing problem of how heat distributes itself through an object over time.

Geiringer's work in 1918 was on *double* trigonometric series, which involves what happens when, instead of adding single sines and cosines together, as Fourier did to construct representations of functions, you add *products* of sines or cosines together. So, instead of something like $a1$ $\sin(x) + b1 \cos(x) + a2 \sin(2x) + b2 \cos(2x) + ...$, Geiringer looked at quantities like $a1 \sin(x)\sin(y) + a2 \sin(2x) \sin(y) + a3 \sin(x)\sin(2y) + ...$, which had been studied previously by Martin Krause and G.H. Hardy. Geiringer provided important criteria for when double trigonometric series do and don't converge, which proved foundational to the evolution of Fourier Series in the twentieth century.

For the two years following her dissertation, Geiringer worked as an assistant editor at the *Jahrbuch über Fortschritte der Mathematik*, before moving on to the University of Berlin in 1921 as research assistant to Richard Edler von Mises, who did research in statistics and probability and their application to problems in fluid mechanics and aerodynamics, among other mathematical interests. With that position, she became the first woman with an academic appointment in mathematics at the University of Berlin, rising in 1927 to the position of *privatdozent* in applied mathematics. She married fellow mathematician Felix Pollaczek in 1921, and gave birth to her only child, Magda, in 1922. Pollaczek and Geiringer separated in 1925, leaving Geiringer to raise Magda on her own while continuing her teaching and research responsibilities, the couple ultimately divorcing in 1932.

Professionally, Geiringer's time at the University of Berlin resulted in arguably her most significant discovery, contained in her 1931 paper, 'Beitrag zum vollständigen ebenen Plastizitätsproblem'. This was the paper that presented a new, and far improved, way of modelling the behaviour of metals under stress, that ultimately led to the development of slip-line theory, which still stands as an important way to visualise and compute what will happen when metals are introduced to different pressures.

What we want to be able to do is to say: when you apply this stress to this metal, this is how it will respond. How they do this in slip-line theory is by creating a net of alpha and beta lines that criss-cross the metal surface in question. The alpha lines show you the directions of maximum stress on the object, and generally speaking, the more bendy they are, the greater the average stress. Here's an expertly rendered example of one such net that was not at all drawn on a napkin:

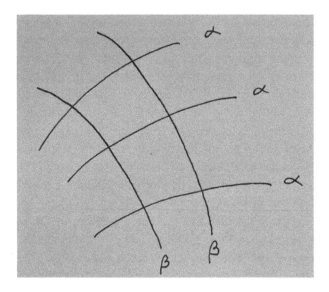

Previously, the Hencky equations had provided a clever way of figuring out the hydrostatic stress at any point on an alpha line, provided that you knew the stress at any other point on that line, but what they didn't do was provide a relatively pleasant way of calculating the velocity field along the surface, i.e. of telling you where the atoms are going to *move* and how violently they will do so. The Geiringer equations do this, relating v, the velocity of a particle at a given point, to the direction of maximum stress there. These equations, while not always simple to solve for a given situation, at the very least represented a marked improvement over the prevailing methods of the time.

It was at this moment, with Geiringer at the top of her mathematical form, that German political developments derailed her career. Hitler's rise to power in 1933 meant that Jewish intellectuals like Geiringer had to scramble for new academic positions outside of Hitler's reach. In 1933, Geiringer fled to Brussels, and ultimately established herself in Turkey in 1934, where the government of Kemal Atatürk was determined to improve the state of Turkish education by importing 200 world-class European professors, among them Geiringer and von Mises. Fortunately, mathematicians require less specialised equipment than, say, a physicist or a chemist, and Geiringer was able to carry on her work, publishing eighteen papers during her four-year Turkish sojourn, including interesting forays into the application of probability to biological problems that extended G.H. Hardy's early work in that field.

The death of Atatürk in 1938 also meant the death of his grand educational initiative, and the imported professors were duly informed that their five-year contracts were not going to be renewed. With Austria consumed in the Anschluss of 1938, and all signs pointing to the full-scale European war that would break out in 1939, Geiringer's best option for a life of peace and research appeared to be the United States, where she and von Mises emigrated in 1939.

The flight of European intellectuals to the United States in the 1930s and 1940s was a boon to American research and university development, but it also meant, in the short term, too few positions available with too many destitute but brilliant professors applying for them. For the rest of her life, Geiringer was unable to achieve a position in the United States commensurate with her abilities. From 1939 to 1944, she taught at Bryn Mawr for low pay. In 1943, she married von Mises, who was teaching at Harvard at the time, but as she was unable to find a university position in Boston, the newly married couple had to live in different states if they wanted to both keep working.

In 1944, Geiringer became the mathematics department chair at Wheaton College, in Illinois, which sounds like a step up until you realise that Wheaton's mathematics department at the time consisted of only two people. Her publication rate slowed, her most significant work in terms of long-term impact being a mimeographed series of lectures on the Geometrical Foundations of Mechanics she delivered in the summer of 1942 at Brown University, which saw wide distribution. In 1953, von Mises died and, as often happens in the history of women in science, Geiringer's life became largely consumed in the task of editing her late husband's works, for which task Harvard made her, at long last, a Research Fellow. In 1958, she completed and published von Mises's *Mathematical Theory of Compressible Fluid Flow*, which led to a collaboration with Alfred Freudenthal that produced her last paper prior to her official retirement, 'The mathematical theories of the inelastic continuum', which represented a continuation of her interest in, and fundamental contributions to, the study of material plasticity.

Geiringer retired in 1959, but continued editing the work of von Mises, publishing his *Mathematical Theory of Probability and Statistics* in 1964 with updated material pulling from the last decade of mathematical advances. Her final decade was spent deepening not only her love of mathematics but her larger interests in literature and music. She died in 1973 of influenzal pneumonia while visiting her brother in California.

FURTHER READING: There isn't a full work on Hilda Geiringer, but you can find bits and pieces of her story here and there. Renate Strohmeier's *Lexikon der Naturwissenschaftlerinnen* (1998) has a nice little column on her, and of course *Women of Mathematics: A Biobibliographic Sourcebook* has a bit more. If you want to get started in the awesome world of Fourier Theory, the book I usually recommend is Stein and Shakarachi's *Fourier Analysis: An Introduction* (2003), which is nice and clearly laid out.

Chapter 16

Fearless Symmetry: Dorothy Wrinch and the Founding of Mathematical Biochemistry

By attempting everything, Dorothy Wrinch ended up accomplishing nothing.

For half a century, this was the standard final verdict on the life and work of mathematician and logician Dorothy Wrinch. Undoubtedly brilliant, so the assessment ran, she squandered her gifts by spreading herself too thin, and as a result by life's end had nothing but a collection of disconnected discoveries and blaring public missteps to show for all her energy, charisma and intelligence.

Had she been a man, her devotion to her pet theories even beyond the point of their debunking would have been shrugged off as an eccentricity that ought not detract from her larger work (as with Newton's fervour for lost ancient wisdom, Kepler's mysticism, or Linus Pauling's obsession with Vitamin C), but as a woman, her refusal to back down was considered so unladylike, so shrewishly unreasonable, that it invalidated everything else she put forth, until by the time of her death, the woman who was a philisophico-mathematical student of Bertrand Russell and an innovator in X-ray crystallographic algorithms was dismissed as a cantankerous crackpot who had paid the just price for her brazenness.

The real Dorothy Wrinch (1894–1976), however, cannot be contained within the bounds of such a simple narrative. Neither so out-of-her-depth and congenitally deluded as her detractors suggest, nor so irrationally and helplessly victimised as her sympathisers would have it, Wrinch was an inventive polymath who scored important technical contributions and pushed conceptual boundaries even as her behaviour in defence of her central theory alienated would-be allies and left her increasingly isolated and disillusioned.

Her father was an engineer, and her mother an ex-school headmistress, and, as might be expected, Dorothy was given access to all the education that the late Victorian era had to offer to a young English girl. She entered

70

school at the age of 4 and made her way steadily through a morality-and-piety heavy curriculum that bore the full imprint of Thomas Arnold's long influence. Her teachers recognised her brilliance but felt that, all the same, she'd come to a bad end. She did, however, leave secondary school with generally positive recommendations and a barrage of mathematical and linguistic certifications, and so made her way to Girton College.

By 1913, when Wrinch arrived at Girton, the college already had a reputation for producing women of academic brilliance. Hertha Ayrton had graduated from Girton in 1880 on her way to making fundamental discoveries in fluid dynamics and perfecting the arc light, as did our friend Charlotte Angas Scott, whose example Wrinch was (not always favourably) held to in her early career. Future botanist Ethel Sargent had graduated in 1885, and 1893 saw the graduation of mathematician Grace Chisholm Young. These women, by their rigorous work and unimpeachable characters, had cleared the way for a new generation of Girton graduates, young women who felt comfortable not only challenging Victorian assumptions about women intellectuals but some of the most fundamental principles of Western civilisation itself.

Wrinch came to Girton as a student of mathematics at a time when that field was experiencing a complete and invigorating re-examination of its capacities and starting principles. In 1899, David Hilbert created a new axiomatisation of geometry on rigorous modern lines that cast away at last two millennia of Euclidean clutter, while from 1910 to 1913, Bertrand Russell and Alfred North Whitehead published the three volumes of the *Principia Mathematica*, an impossibly dense but bracing attempt to recreate the entirety of mathematics itself from fundamental principles of logic. Logic was maths, and maths was everything, and Wrinch was swept up in the wave of mathematical optimism.

During her first few years of college, Wrinch was being pulled both intellectually and sympathetically into the orbit of Bertrand Russell. His logical genius appealed to her sense of mathematics' sure and eternal power, his outspoken and solitary stance against the First World War reassured her that her own anti-war stance was not shameful or cowardly, and his brilliant religious scepticism accelerated her own tendencies towards non-belief in the variously oppressive trappings of organised religion. If ever two people were destined to meet and become lifelong friends, it was Dorothy Wrinch and Bertrand Russell, and so they did, in 1916.

By that point, Wrinch had experienced a devastatingly low result on her first Tripos examination as a result of her insistence on taking the exam after one year of study instead of the usual two, but managed to score towards the top of the pack in the second round, while simultaneously studying the leading lights of early twentieth-century philosophy. When Wrinch met Russell, they experienced an immediate meeting of minds. Russell was impressed by her knowledge of, and enthusiasm for, the deep mysteries and puzzles posed by mathematical logic, and included her in the small circle of trusted friends who worked together during the war on mathematical problems even as Russell was jailed for his anti-war statements.

In what was to become a pattern for Wrinch, she devoted the full force of her fertile mind in the name of a dying but beautiful cause. She worked away at problems of logic that bedevilled Russell, diving deeper and deeper into the intoxicating complexity of logic's infamously impenetrable notation, all in the name of at last constructing a perfectly consistent and self-contained theory of mathematics.

A decade and a half later, in 1931, Kurt Gödel would publish his Incompleteness Theorems, effectively proving that axiomatic systems are inherently limited, and therefore that programmes like Hilbert's and Russell's are doomed to ultimate failure. Fortunately, by then Wrinch had already moved beyond logic as new fields caught her fancy. Under a pseudonym, she published *Retreat from Parenthood* in 1930, a book laying out her scientific system of parenting which called for a new government department to provide organised support for new parents, allowing women more freedom to work, and increased opportunities for restorative leisure. She tackled the emerging big questions in the history of science about how ideas get accepted and promoted in scientific communities. And, lest we think that she operated purely on the most impossibly vast dimensions of thought, she also made solid contributions to the technical but important theory of trigonometric series.

All very well, some mathematicians thought, but what does it all add up to? G.H. Hardy, one of the most important mathematicians of his, or any, age (and also the promoter of Indian mega-genius Srinivasa Ramanujan), was a devoted feminist and promoter of women in mathematics, and even he struggled to understand where Wrinch was going with her multifarious interests. In his opinion, she had perilously divided her attentions, and as such was achieving far fewer results than she ought, and couldn't be ranked in the same tier as Charlotte Scott or Grace Young.

All the same, he attempted to help her secure funding, only to be frustrated by the discovery that, while she was telling *him* she was ready to settle down and concentrate fully on mathematics if only he could help her get a grant, she was telling somebody else that she was ready to settle down and concentrate fully on social institutional theory, if only *they* could help her get a grant. Today, the idea of an academic pursuing widely different fields simultaneously, and seeking funding for each, has lost its ability to shock, but in the pre-war years, a mathematician seeking funding was expected to be a lifelong devotee of their field of study. Why give a grant, after all, to somebody who was only going to keep half an eye on her proposed field, when so many pure and committed researchers were waiting in the wings? What today we would laud as a devotion to multidisciplinary research appeared at the time as fickle indecision, and Wrinch would have to wait decades for scientific styles to catch up with her own ways of working and thinking.

In the meantime, in the 1930s Wrinch hit at last upon the field that would claim her more or less undivided interest for the remaining four decades of her life – the intersection of mathematics with biology, and in particular the problem of the structure of proteins. To understand this part of the story, we have to put ourselves back in the mindset of 1935, before the structure of DNA was known, before every biology textbook in the world carried pictures of chains of amino acids folding at the behest of hydrogen bonds and disulphide bridges to form intricate three-dimensional structures known as proteins. At that time, though many were convinced that the key to proteins lay in their structure as long chains of amino acids linked by peptide bonds, others were not so sure. The variety of physical and chemical characteristics of proteins, the sceptics argued, surely couldn't be the result of a structure as simple and mono-dimensional as a chain.

Wrinch proposed to bring mathematics to the problem – what sorts of geometrical structures fit the given data and had enough variety to explain all the behaviours demonstrated by proteins? Wrinch had entered the field of biology by proposing a genetic structure theorising that chromosomes were essentially fabrics made up of a protein warp and a nucleic acid weft that meet up at the edges to form a tube. It was ultimately nowhere near correct physically (though it did correctly posit that the linear sequence of amino acids is important in a gene's function), but it was stimulating and elegant, and it demonstrated what might be done when a mathematician and geometer brought her unique perspective to a biological problem.

It was time, then, to enter the debate about proteins and here, as in the case with logic, Wrinch had the misfortune of entering the field when some highly attractive but ultimately incorrect measurements and principles were widely believed. In particular, it was held that proteins consist of amino acid residues whose number is always a product of a power of 2 and a power of 3, and the number of molecular weight classes that proteins belong to is sharply limited. These notions, both of which are entirely wrong but were generally accepted, were seized upon by Wrinch as the mathematical basis for her protein models. Instead of linking up amino acids in long chains, she suggested in 1936 that they formed elegant hexagonal fabrics which could, in turn, form three-dimensional Platonic cages. She called this the 'cyclol' structure for proteins, and it created a firestorm of controversy, publicity, and, most importantly, research, as some raced to prove it and others to show that it was utterly impossible.

It was attractive in that it explained in terms of symmetry and structure why the amino acids should show up as powers of 2 and 3, and also carried a physical interpretation for how protein denaturing worked (the cage simply reverting to its two-dimensional fabric state, which carries a different chemistry). It was a beautiful and seductive model, sporting symmetry and mathematical substance, but it was vehemently opposed almost from the start, and no voice against it was more influential than that of chemistry's reigning monarch, Linus Pauling.

Pauling saw several immediate problems with the cyclol model, and when he had the chance to question Wrinch about them directly, was so disturbed by her unwillingness to engage with the nuts and bolts chemistry of her theory that he made it a point to publicly weigh in on the matter with a crushing critique. He pointed out that the bond energies of the cage model were too high to be nature's preferred structure, that the density of a cyclol cage didn't match the density of known globular proteins, that the covalent bonds she proposed as holding together her cage couldn't be undone easily enough to account for the regularity of protein denaturisation as effectively as a hydrogen bond hypothesis, and that her 288 ($2^5 * 3^2$) amino acid base unit didn't faithfully reflect the diversity of known proteins.

Pauling's 1938 paper listing what he believed were the most glaring faults of the cyclol model all but sealed its fate as far as the larger scientific community was concerned. But Wrinch was not convinced. She wrote to journals pointing out how some of Pauling's calculations were flawed (they were, but not enough to change the evidence by enough to warrant a reconsideration of the cyclol model), requested that chemists she knew personally devote time to performing experiments that might prove her correct, and when those chemists ultimately grew tired of her unwillingness to accept their results when those results went counter to the cyclol model, asked for funding to hire her own assistant to do the experiments herself.

The subsequent years were filled with bitterness and controversy. She sparred unsuccessfully with crystallography legend Dorothy Hodgkin, insisting that insulin displayed a cyclol structure, compelling Hodgkin to definitively remark that there was no evidence for such a claim. Colleagues expressed frustration whenever she waived away their chemical objections as unimportant details not relevant to the larger theoretical picture. There was a scandal when she redirected her second husband's research funding to pay for her own experiments, and her last book, *Chemical Aspects of Polypeptide Chains* (1965), continued to argue for a cyclol model nearly three decades after Pauling's paper, prompting one reviewer to comment, 'If a reviewer had been asked to consider a book on the "phlogiston" theory years after Lavoisier had shown that it was incorrect, he might have felt the same way as I have with this one.'

These were hard years, yes, but not barren ones. Though, to the scientific community, Wrinch was That Sad Cyclol Obsessed Lady, and though her increasingly desperate unwillingness to admit defeat certainly added to that impression, these were also the years that she was quietly working a potential revolution in yet another field of science: X-ray crystallography. She brought her knowledge of mathematics and topology to the labour-intensive process of interpreting Patterson maps to intuit a crystal's molecular structure from the pattern obtained when X-rays are shone through it. Her algorithm was technically correct but was, unfortunately, not of immediate use for the protein crystallography of its time, lacking as it did the computing power to perform the underlying computational gymnastics. Decades later, the result would be rediscovered and widely applied in the computerised interpretation of molecular structures.

At this point it is hard not to feel a bit like Hardy did almost a century ago: what are we to make of all this? What does Dorothy Wrinch's life amount to? Must we tie her reputation entirely to the success or lack thereof of the cyclol model, as she herself did, or can we see her as something more than that one beautiful hypothesis and its disastrous aftermath? Certainly, we are in a position now to understand her better than her contemporaries ever could. She contributed to philosophy, the history of science, geometry, molecular biology and chemistry, and was the first in a wave of talented multidisciplinary thinkers to bring the advanced topological tools of mathematics to the realm of biology. Her ideas, even when wrong, spurred people to develop and apply new techniques to their study, and her lectures were roundly reported as inspiring master-classes that taught a generation to possess the insight of a chemist and the eye of a geometer simultaneously. She spoke out against destructive and unnecessary war even as her home nation intoxicated itself on jingoistic euphoria, pressed for reform in governmental policies towards mothers, and held religion accountable for its manifold intellectual and spiritual abuses, and did all of that WHILE acting as an interdisciplinary scientific pioneer.

Dorothy Wrinch's only daughter died in a fire in November 1975. Dorothy lived for three more months thereafter, unspeaking, and died herself in February 1976 after five years of retirement from a career that had brought pain, and beauty, though who can say in what ratio.

FURTHER READING: The major book about Wrinch's life and science is 2012's *I Died for Beauty: Dorothy Wrinch and the Cultures of Science* by Marjorie Senechal. It brings Russell's intellectual companions and those of the Vienna Circle to robust life, and does well in explaining the points of contact between these schools and Wrinch. It is written in a loose and artistic style that definitely isn't the norm in scientific biography, culminating in a proposed libretto for an opera about Pauling and Wrinch that takes up a full chapter. Her characterisation of Pauling is, perhaps not surprisingly, not entirely balanced, for which Thomas Hager's *Force of Nature: The Life of Linus Pauling* (1995) provides a corrective – he has a dozen pages devoted to the cyclol controversy where the portrayal of Wrinch is, perhaps not surprisingly, not entirely balanced. Between the two books, one comes out at a happy and probably accurate medium wherein Pauling's concerns were largely legitimate and Wrinch's theory, when you consider the reigning assumptions it was nested in, was not nearly as improbable as it sounds now.

Chapter 17

In Defence of the Soil: One Century with Hydrodynamic Mathematician Pelageya Polubarinova-Kochina

Water is that great, terrible thing. Its chemical properties make it a magnificent solvent and coolant, which is wonderful if you're trying to build a multicellular organism, but mathematically its behaviour was for centuries the stuff of scientists' most frenzied nightmares. Unlike solids, which more or less sit there, and gases, which have molecules so far apart that we can usually make a set of ideal assumptions that allow us to make good approximations as to their behaviour, liquids live in this nebulous middle world of mathematical complexity, and *water*, with its powerful intermolecular forces, is a particularly nasty customer. For ages, we didn't *need* to know all that much about water's more complicated aspects, so we didn't much bother plumbing those depths, but with the coming of the modern age and its craze for large-scale building, suddenly the need to deeply understand the motion of water through soil, and what it does to that soil's long-term stability, became a pressing concern.

Some motions towards modelling the behaviour of water mathematically were begun in the mid- to late nineteenth century by the likes of Darcy, Dupuit, Forchheimer, Lembke and Zjoukovsky, but it was the early twentieth century that saw the subject burst into its full rigorous maturity under the watch of two brilliant Russian mathematicians, a married couple devoted to the Soviet ideal that academic knowledge was at its most valuable when applied to problems that improved the lives of everybody: N.E. Kochin and Pelageya Polubarinova-Kochina (1899–1999).

Pelageya Yakovlevna Polubarinova was born when the Russian tsar still sat confidently on his throne, completely unaware that in two decades he would be removed from power and murdered by his own people, and she died eight years after the fall of that Soviet Union which did the murdering. In between, she was one of her country's most honoured mathematical minds, who made fundamental discoveries into the application of differential equations to modelling the behaviour of groundwater movement, and who at the age of 60 uprooted her life to bring that knowledge to a harsh and unforgiving new climate to the benefit of those living there.

Everything about the setting of Polubarinova's story is on the grandest of scales – she worked on towering projects that reshaped the destinies of people against the backdrop of revolution, Stalinism and a nation's lurching will to achieve the future. For all of the dramatic bigness surrounding her, however, we have all too little to say about the early years of Polubarinova herself. In 1974, an authorised memoir of her life was published, but as a party-sanctioned account of one of the country's most valuable intellectual figures, its authenticity on many points has been called into question, and it is more often useful for what it doesn't say than for what it does.

What we seem to know is that Polubarinova was born in 1899 in Astrakhan, one of Yakov Polubarinov and Anisiya Panteleimonovna's four children. Sometime in her youth, the family moved to St Petersburg, where she enjoyed the more rigorous mathematical options on offer in what was then Russia's capital city. In 1918, her father died, leaving her as the primary bread-winner for her mother and two of her siblings (one of her sisters died of tuberculosis during the Civil War that followed the Bolshevik Revolution). These were years of toil and hardship, as Polubarinova worked at the Main Physical Laboratory to earn money, while at the same time attending Petrograd University to further her studies, pushing through all the responsibility and work even when she contracted tuberculosis herself.

Personally challenging, these were nevertheless years of intellectual excitement as Polubarinova benefited from the Soviet Union's policy of allowing equal access to upper division science topics for everyone, regardless of gender. She was able to take classes in hydrodynamics, complex variables, differential equations and theoretical meteorology that would have been off limits to her just five years earlier. Graduating in 1921, she went on to work for Alexander Friedmann, known today for the Friedmann equations which predicted that the universe is expanding, but who was then mostly known as a theoretical meteorologist.

Working in the Division of Theoretical Meteorology, she performed computations and met mathematician Nikolai Evgrafovich Kochin, who would become her husband in 1925. Their marriage featured a telling simplicity – they attended a brief civil ceremony, took their guests out to tea, and then got back on with the work before them. Nikolai was devoted to the ideals of Communism, and had served in the Red Army during the Civil War, and like Pelageya he was gripped by the idea that those with mathematical talents should use those skills whenever possible for the greater good. As

Nikolai rose through the ranks of the Soviet academic world, Pelageya devoted herself to teaching mathematics to the workers who now had access to the Russian educational system, serving as an instructor at the Institute of Transportation from 1925 to 1931, the Institute of Civil Aviation Engineering from 1931 to 1934, and as a professor at Leningrad University after 1934.

Pelageya found her work as a professor far less demanding than her duties as an instructor had been, and resolved to complete her PhD, which she began in 1934 and completed in 1940, becoming only the third Russian woman to receive the degree of Doctor of Physical and Mathematical Sciences. Her work was increasingly centred upon the problem of filtration, of creating a mathematical model for how water moved underground that could be used when planning large-scale construction or hydroelectric projects. Some progress had been made on this front before, but an equation that was able to represent, for any point in a given section of subsoil subject to certain initial conditions, what the potential for water movement might be, had eluded the world's best minds.

Polubarinova-Kochina attacked this problem using the full artillery of her mathematical upbringing, employing her knowledge of differential equations and complex variables to craft functions that modelled two-dimensional steady flow of underground water under increasingly complicated circumstances. Beginning with her classic 'Application of the theory of linear differential equations to some problems of groundwater flow' (1938), through 'On unsteady groundwater flow in two layers of different density' (1940), 'On unsteady flows in the theory of seepage' (1945), and 'On unsteady groundwater flow in seepage from reservoirs' (1949), and culminating in her landmark textbook *Theory of Groundwater Movement* (1952), Polubarinova-Kochina tackled and solved some of the greatest engineering problems involving the interface of soil and water, paving the way for her greatest adventure in 1959.

At 60 years old, Polubaroniva-Kochina was one of those rarest of creatures: a scientist who had risen to prominence during the Stalin era and maintained her position not only throughout but beyond. The value of her work was so beyond dispute, her commitment to the cause of science grounded by human need so apparent, that the academic purges which withered so many other intellectuals during Stalin's regime did not touch her. (Or, at least, they don't seem to have – her memoirs are conspicuously silent on the subject of Stalin.) By 1958, Stalin had been dead for five years and Soviet science had largely picked up the pieces of the destruction he had wrought and was marching forward to new projects. That year, Polubarinova-Kochina was invited to be one of a small group of scientists who would uproot their lives and move to Siberia to attempt to bring hydroelectricity and other infrastructure into that desolate and forbidding corner of the world.

At that point in life, Polubarinova-Kochina had nothing to prove. She was recognised the world over as an expert in her field of applied mathematics. She had a good and respectable position in Moscow, where she lived close to her children and grandchildren (her husband had died in 1944). And yet, at the age of 60, she gave all of that up to live for more than a decade at the bleak edge of Soviet civilisation, heading the department of applied hydrodynamics at Novosibirsk from 1959 to 1970, applying new mathematical tricks to the problems presented by irrigating and electrifying new and barren lands.

Returning home at last at the age of 71 she continued her mathematical research while the Soviet Union lavished titles and celebrations upon her. She had already been made an Academician of the Academy of Sciences in 1958 and received four Orders of Lenin. On the occasion of her seventieth birthday, she was made a Hero of Socialist Labor, and on that of her eightieth, she was given the Order of Friendship of Peoples, while new appreciations of her life and work were published on her seventieth, seventy-fifth, and eightieth birthdays. Her last paper, 'Some properties of a linear-fractional transformation', was published in 1999, shortly before her death at the age of 100.

There are parts of Pelageya Yakovlevna Polubarinova-Kochina's life we shall probably never know, tales of anxiety in the darkest of times, and struggles on civilisation's thin edge, but what we have, what we will always have, is her eight decades of steady and dedicated work to apply her age's best mathematical insights and techniques to the solving of its most intractable engineering problems. There are those who have power, and water, and a place to live, thanks to her seemingly prosaic work unravelling the complex interface between land and liquid. She lived in that brisk moment when practice, society and intellectual effort all marched with the same gait, and provided one of its most stirring examples of mind and conscience, pure raw brainpower and idealistic commitment, harnessed together in the name of public work and the general good.

FURTHER READING: Polubarinova-Kochina's *Theory of Groundwater Movement* was translated by J.M. Roger de Weist in 1962 and features a couple of introductory pages by the translator exploring her place in the history of hydrodynamics, and two of the chapters prominently feature the results of her own research (Chapters 7 and 14), but unless you are really up on your differential equations it is probably going to be rough going.

Chapter 18

Letting Loose the Dogs of Chaos: Mary Lucy Cartwright's Pioneering Portrayals of Functions Behaving Badly

Our concept of living in a universe with a knowable and predictable future has taken two stunning blows in the last century, first from quantum mechanics in the 1920s, which uncovered a number of quantities which don't play well together and place limits on the degree to which we can measure the world around us, and then, with far less fanfare but with potentially even greater repercussions, from some odd mathematical results discovered by Mary Lucy Cartwright (1900–1998) and J.E. Littlewood while investigating the behaviour of radar signal amplifiers during the Second World War. Their results, worked out in rigorous detail over the next decade, were the precursors of an entirely new approach to what science could and could not predict, a field today known as Chaos Mathematics.

For most of us, our knowledge of chaos mathematics more or less begins and ends with Jeff Goldblum's speech while sprinkling droplets of water on Laura Dern's hand in *Jurassic Park* – minute unobservable changes can add up in ways that make predicting macroscopic events impossible. We might even know the 'butterfly effect' of Edward Lorenz's thought experiment – 'If a butterfly flaps its wings in Brazil, could it cause a tornado in Texas?' The situation that presented itself to Cartwright in 1938, however, was decidedly less poetic, and entered not on the wings of a lovely insect or through Goldblum's dulcet tones but via misbehaving radar equipment.

With the Second World War blooming in red ruin upon the horizon, England was busily employed attempting to get its radar technology to a consistently accurate standard. Aeroplanes, being made out of metal, are particularly good at reflecting radio waves that are sent in their direction. One could, by sending out a radio wave, and measuring the frequency change of the returning reflected wave, come to a conclusion about where an enemy plane is, and where it is heading. The problem was that reflected signals tended to be very weak, and needed amplification in order to be detected by operators. During the 1930s, however, operators began discovering that, when they required high degrees of signal enhancement, their amplifiers started going wacky, transforming the uniform input signals they were receiving into a cacophony of unpredictable, multi-frequency outputs that seemed to follow no pattern.

There were two ways to solve this issue, one was just to keep brute forcing the problem until a configuration was found that produced reliable results, but there was no telling how long that would take. The other was to send the problem off to mathematicians, and see if they could, by analysing the mathematical properties of amplifier systems, come up with an explanation of the phenomenon and therefore potentially a targeted solution. In 1938, the problem fell in the laps of Cartwright and Littlewood, two function theorists who seemed an unlikely pair to tackle such an applied problem, but who turned out to be exquisitely well tailored for the job.

Mary Lucy Cartwright was born on 17 December 1900 in the village of Aynho, in west Northamptonshire, to a family of distinguished lineage, including the poet John Donne and Royal Society curator (and planetarium inventor) John Theophilius Desaguiliers as ancestors. Her father was a rector, and the family had the means to provide young Cartwright with governesses until the age of 11, when she began attending Leamington High School, where she stayed with relatives during the school week, returning home for weekends. She remained at Leamington from 1912 to 1915, then was a boarding student at Graveley Manor School for a year, before having the course of her life changed at Godolphin School in Salisbury, which she attended from 1916 to 1919.

At Godolphin, she was taken under the wing of a teacher who had taught herself the higher mathematics that were often omitted from a young woman's education at even the finest of high schools. That teacher, a Miss Hancock, saw Cartwright's raw talent and took active interest in the 16-year-old student, teaching her calculus, analytic geometry, and, most crucially for her future career, the theory of function convergence. In 1919, armed with a good background in pure mathematics, but hardly any in applied maths, Cartwright began attending St Hugh's College at Oxford, where she quickly found her way into the classroom of one of the century's great pure mathematicians, G.H. Hardy. She attended his late-night evening classes, where students would discuss interesting mathematical topics from 8:45 to 11 p.m. Here she truly caught the bug for function analysis that would lie at the centre of so much of her work in the coming decades.

Graduating in 1923, Cartwright turned to teaching, first in Worcester then Buckinghamshire, but by 1927 she was back at Oxford as a graduate student, with Hardy as her advisor. Her work of the late 1920s centred on Abel summation, which is worth talking about because elements of this early work will reappear in her more famous later papers. Suppose you have a list of numbers that follows a pattern that you are able to write down. For example, the list 1, ½, ⅓, ¼, ⅕, ... can be written as following the pattern $1/n$. The question becomes, if I want to add up all the terms of that pattern, onwards to infinity, will they sum to a finite number, or will the sum just keep getting bigger and bigger? Depending on what the pattern is, these questions can either be pretty easy to answer or very, very tricky. Abel summation is a way of finding the sum of a pattern by representing that sum as a related integral, a mathematical object we have any number of techniques for evaluating.

What Cartwright was developing in this early work was a way of thinking that bridges the spaces between the behaviour of integrable functions and the summability of the patterns related to them. She was to expand this type of thinking in the 1930s with her work on function theory and particularly with integral functions. She was interested in questions about the behaviour of different types of functions restricted to a given region of space (often a circle of radius 1 centred at the origin known as the 'unit circle'). Suppose, for example, you and a friend start at two different locations on the edge of that circle, and weave your way however you want towards the same point, and at every step along the way you evaluate an equation, which takes as its input the coordinate of the location you happen to be standing on – what can you say about the outputs of those equations obtained by you and your friend as you both get closer and closer to the shared end point of your journeys?

Cartwright's work of this era, building off earlier efforts by Finnish topologist Ernst Lindelof and her colleague J.E. Littlewood, focused on establishing boundaries on the behaviour of functions defined on different spaces under different restrictions. If you have a box in your hand that displays different numbers as you walk through a space, which are arrived at through an equation located in the box, how do restrictions on the equation in the box place bounds on the numbers on the display, or determine how many times that number will come up zero? Cartwright was interested in finding increasingly precise and efficient ways of ensuring that that number on the box display will not blow up to infinity no matter where you go in the space. She famously established conditions for the equation inside the box under which, if you saw that the displayed number was under a certain value for any integer valued input, you could then conclude that it would also be under that value for any real number valued input. So, if the box is behaving when you input 1, 2, 3, or 500, you know it will also behave if you plug in 1.8, or the square root of two, or 5 11/13, which brings answering questions about real number functions down to answering questions about behaviour at the integers, which are generally far more pleasant to answer.

In 1930, Cartwright received her doctorate, and began her association at Girton College, which was to last for her entire career, serving first as a research fellow, then college lecturer, and by 1936 as Director of Studies for Mathematics. It was while in this position that the radar problem was dropped in her lap. Seemingly, nothing could be further from the field of expertise of an abstract function theorist than an applied question springing from electrical engineering, but in truth Cartwright's background was perfect for cutting through the conundrum. By 1938, she had been grappling for a decade with problems about the conditions required for functions to behave reasonably on given spaces, and the radar problem could be viewed from precisely this perspective.

In 1927, a Dutch electrical engineer named Balthasar van der Pol had noticed while working with vacuum tubes that when they were driven at certain frequencies, irregularities could be heard in the output signal. He couldn't know it yet, but those irregularities were the result of chaos effects that lay deep in the structure of the mathematical formula he created to describe the waves produced by his vacuum tubes. That formula was a differential equation, $x'' - k(1-x^2)x' + x = bk \cos(\lambda t)$, where the right side of the equation describes the presence of external disturbing factors, and $x(t)$ describes the position of the wave at any given moment in time, with x' and x'' being the first and second derivatives of that function.

Solving a differential equation involves finding an equation $x(t)$ which meets the given criteria. So here, for example, Cartwright's challenge was, given values of k, b and λ, to find an equation such that, when you take the second derivative of it at any moment in time, and then subtract off k times the first derivative of it at that time multiplied by 1 minus its own value squared, and THEN add its own value, you end up with $bk \cos(\lambda t)$ at that time. It is a daunting task at the best of times, which Cartwright and Littlewood approached by investigating a more general form of the van der Pol equation, doing what they did best – establishing parameters for good behaviour.

They considered the general differential equation $x'' + kf(x,x') + g(x,k) = p1(t) + k\, p2(t)$ and wondered what would happen as you started tinkering with each part. What they found was that, as long as $g(x,k)$ is relatively

Catherine de Parthenay's promising mathematical career was cut short by the French Civil Wars, which ended in the destruction of her family estate and exile of her sons. (Wikimedia Commons)

Émilie du Châtelet brought Newtonian physics to France through her translations and commentaries on his work. (Wikimedia Commons)

Maria Gaetana Agnesi was a polylingual child prodigy whose textbook on differential and integral calculus was the pinnacle of the pure geometric approach to those topics. (Wikimedia Commons)

Mary Somerville wrote some of the nineteenth century's most influential books on astronomy, physics, and mathematics, which were responsible for reinvigorating British mathematics with new continental ideas. (Wikimedia Commons)

Sofia Kovalevskaya was both a novelist and one of the nineteenth century's greatest minds in the area of differential equations, who died young at the hands of influenza. (Wikimedia Commons)

Charlotte Angas Scott, the 'Englishwoman in America', brought continental techniques and organisation to United States mathematics, helping to establish the American Mathematical Society. (Wikimedia Commons)

Emmy Noether created one of the most important theorems in physics, unifying all possible laws of conservation into one master expression, and was a key figure in linear algebra. (Wikimedia Commons)

Author of over 300 papers, Olga Taussky-Todd was the central figure behind the mid-twentieth century push to more deeply investigate the properties of matrices. (Wikimedia Commons)

Dame Mary Cartwright's 1945 paper on non-linear differential equations was a foundational work in the history of chaos theory. (Wikimedia Commons)

Marjorie Lee Browne was the third black woman to earn a mathematics PhD in the United States. She studied how best to use computing technology to improve numerical analysis. (Wikimedia Commons)

Katherine Johnson
co-wrote NASA's
first internal text
on the mathematics
of space flight,
and computed
the trajectories of
the Mercury and
Apollo missions.
(Wikimedia
Commons)

Julia Robinson
was the first
woman president
of the American
Mathematical Society,
and the solver of
David Hilbert's
famous Tenth
Problem. (Wikimedia
Commons)

Karen Uhlenbeck was the first woman to win the Abel Prize, and her work analysing Yang-Mills equations in four dimensions laid the ground for Donaldson's later Fields Medal winning work. (Wikimedia Commons)

Raman Parimala surprised the mathematical community at a young age with her discovery of a non-trivial quadratic space over an affine plane. (Wikimedia Commons)

Maryam Mirzakhani was the first woman to win a Fields Medal, and a daring investigator into the properties of Teichmüller spaces, before her early death at the age of 40. (Wikimedia Commons)

Fan Chung is an expert in the application of combinatorics theory to the evolution of large systems, such as those encountered on the Internet. (Wikimedia Commons)

linear (i.e. the output is regularly proportional to the input), and as long as k is large, then the solution $x(t)$ to the differential equation converges to a single result over time, but that when you start to move away from linear g's and large k's, convergence is by no means guaranteed, and in fact can produce numbers of different possible waves of uncertain stability.

Cartwright and Littlewood's analysis of the radar problem took seven years to work out, by which time engineers had produced their own solution to the amplification problem using trial and error methods, but the resulting paper, 1945's 'On non-linear differential equations of the second order I' was a tour de force that the pair would expand on throughout the 1940s, and which played no little part in Cartwright's election as a Fellow of the Royal Society in 1947, followed closely by her promotion to Mistress of Girton College in 1948, a position she held until 1968, shepherding the institution smoothly through an era of considerable social change.

When she retired from Girton in 1968, Cartwright spent the next three years visiting foreign universities before settling down in 1971 to her lodgings at 38 Sherlock Close in Cambridge where she spent the next two decades of her life, continuing her mathematical researches, and producing her final paper, a return to the Van der Pol equation, in 1987. She had lived her days with an efficiency and emotional reserve that allowed her to accomplish much as both an administrator and researcher yet also kept most colleagues and students at a somewhat professional distance. To the mathematical community, she was known and respected for her exquisite results on how functions behave near singularities, the prerequisites for uniform function behaviour, and on how to employ conformal mappings to establish limits on the roots and asymptotic values of functions, but her pioneering role in detailing the mathematical behaviour of a chaotic system was buried in such an intimidating wall of functional reasoning that it failed to capture the public imagination in the same way that the visually arresting sets of Mandelbrot or the accessible explanations of Lorenz did, causing this pioneer in the field of chaos mathematics to often be either omitted entirely from its history or relegated to a (sometimes very literal) footnote. Fortunately, that situation is beginning to change, and the road to Jeff Goldblum is increasingly placed less in the flutter of a butterfly's wings and more in the squelch of a vacuum tube, straining to have its subversive deviations understood by a world not yet ready for them.

FURTHER READING: There are two very nice memoirs of Cartwright's life that are relatively accessible, the first being W.K. Hayman's Royal Society memoir upon her death, which is available online, and the second an appreciation by none other than Freeman Dyson in the pages of the wonderful collection *Out of the Shadows: Contributions of Twentieth-Century Women to Physics* (2006) edited by Nina Byers and Gary Williams. Of course, one would think that, given the importance of her analysis of the Van der Pol equation, she would be prominently featured in what most people consider *the* go-to popular account of chaos theory, James Gleick's 1987 book *Chaos: Making a New Science*, but she appears in that volume as only a mention stuffed into an endnote, while Gleick focuses more on the work of Smale and Lorenz as the founders of chaos theory. Maybe this will change in future printings, but as of the twenty-sixth printing, it is still the case.

Chapter 19

Equilibrium States: Tatyana Alexeyevna Ehrenfest-Afanassjewa and the Development of Statistical Mechanics

Whereas few European scientists escaped the politico-intellectual gnash of the 1930s unscathed, arguably none faced quite the looming combination of compound miseries thrust at a single, seemingly unbreakable individual of that era, Tatyana Ehrenfest-Afanassjewa (1876–1964). Displaced by systemic anti-Semitism, perpetually lacking stable work, shouldering the depression of her husband that led to his suicide and to the murder of one of her children, and bearing up against the freak death of another, Ehrenfest could easily have sunk under the weight of such dense misfortune, but amazingly she did not, and survived another quarter century after Europe's unrelenting assault upon her person, and not only survived, but formed a wholly new approach to mathematical education and retained her position as a world expert on the mathematical processes underlying thermodynamic change.

Tatyana Afanassjewa was born on 19 November 1876 in Kiev, but at the age of 2, her railway engineer father, Alexey Afanassjew, was committed to a mental asylum following a breakdown, and her mother, Yekaterina Ivanova, took young Tatyana to live with her childless aunt and uncle in St Petersburg. The same year of her arrival in St Petersburg, 1878, saw the opening of the Higher Women Courses (otherwise known as the Women's University) in that city, a concession by Tsar Alexander II, who wished to reverse the trend of Russian women going abroad in search of university degrees that he unwittingly set off with his 1862 prohibition on women's higher education in Russia, and who noted the general success of his more targeted 1873 opening of the Women's Medical Courses in the city.

Whereas it would have made entire sense for a girl of Tatyana's evident gifts to have attended the Higher Women Courses, her guardians decided to place her instead in a teacher training programme, from which she graduated in 1897. Shortly thereafter, Tatyana's uncle died, leaving her free to attend the Women's University, where for the next three years she honed her skills in mathematics and physics. In 1902, she took the momentous step of transferring to the University of Göttingen, which we have met multiple times now as the mathematical mecca featuring the dual attraction of superstars David Hilbert and Felix Klein, and where a young man named Paul Ehrenfest strenuously argued that she ought to be allowed to join the students' mathematical club in spite of her gender. Ehrenfest, who was Jewish, and Afanassjewa, who was Russian Orthodox, soon fell in love, and married in Vienna in 1904 after Ehrenfest received his doctorate from the University of Vienna.

The couple had to renounce their individual religions in order to marry, a fact that held them back professionally, especially in a time when lack of a professed religion was associated with radical movements like Communism and Anarchism. Paul and Tatyana moved to St Petersburg in 1907, where they struggled to find work, but where their home served as the base of an influential informal physics colloquium, dedicated to sharing and discussing the most recent developments in mathematics. In 1910, the couple's second daughter, Anna, was born (the first, a future mathematician also named Tatyana, was born in 1905), adding extra financial pressure to the family, compelling Tatyana to take up a variety of odd jobs, including as a grammar school mathematics teacher, and as a member of the Pedagogical Museum of the Military Academy work group dedicated to formulating new approaches to mathematics education in Russia.

Though difficult financially, this era in their early joint work was fruitful intellectually, and in 1907 they published one of their most enduring early contributions to thermodynamics theory, the Ehrenfest Model, sometimes known as the 'dog-flea model'. In the model, a container filled with gas molecules is connected to an empty vessel. The Ehrenfests investigated the long-term expected behaviour of this system, given different values for the probability q that, at any given moment, a particle should 'choose' to hop from one container to the other. What the model found was that, though choosing higher values of q causes the system to reach its equilibrium state more quickly, it does not impact what that equilibrium state ultimately is. An important result of this model is that it provides estimates for how long it takes the system, if it finds itself in a state that is not the equilibrium state, to return to equilibrium, and for how long you would have to wait for the system to randomly re-assume its starting configuration (spoilers: it is a long time, like twelve orders of magnitude greater than the current age of the universe long). This was important, because it tied into the era's debates about the validity of Ludwig Boltzmann's H-Theorem, which predicted that systems tend generally towards a certain optimal distribution of energy states over time, with deviations from that state brought back under control in a manner similar to the Ehrenfest Model's predictable return to equilibrium.

The next gem of the early Ehrenfest partnership came in 1912 when Paul and Tatyana finally published the survey of statistical mechanics that Felix Klein had first requested of them in 1906 (largely on the strength of the Ehrenfest Model, which they had presented that year). Arising out of Boltzmann, Maxwell and Gibbs's investigations into entropy and energy distributions over time, statistical mechanics is the branch of mathematics that uses the principles of statistics and probability to explain the behaviour of large collections of atoms, molecules, or microscopic substances. 'The Conceptual Foundations of the Statistical Approach to Physics' was a landmark and deeply influential publication that clearly set out the developments in the field thus far, laying out the merits of Boltzmann's H-Theorem and warding off some of the more reckless objections to its results, while at the same time acknowledging the limitations in application caused by some of its more controversial assumptions (such as the 'ergodic theorem' that each particle in a gas visits every location in its container, given enough time to wander). Clear in its writing and comprehensive in its treatment, *Conceptual Foundations* laid out areas for further work and established the foundation for the theoretical study of irreversible processes.

Conceptual Foundations so impressed University of Leiden professor H.A. Lorentz that he decided, upon his retirement in 1912, his seat should go to Paul Ehrenfest, rather than the individual he had previously wanted to attract, Albert Einstein. This was a massive stroke of luck, as Ehrenfest's Jewish ancestry and publicly confessed atheism had both steadily worked against him in professional academic circles for years. Tatyana, however, while recognising the financial necessity of the move, was frustrated that her work in Russian education reform was being uprooted just as she was set to receive the results of a massive survey sent to Russia's university students asking them what their impressions of the strengths and weaknesses of their mathematical training in Russia were. While pedagogical theorists debated the relative merits of abstract proofs versus concrete examples, Tatyana had hit upon the radical notion of actually asking the students who went through the system what they thought about it, and the move to Leiden seemed likely to disrupt this promising work.

While Paul was teaching in Leiden, Tatyana was continuing both her mathematical research and her studies of educational reform, working around the birth of two more children, Paul (1915), and Vassily (1918). Vassily was born with Down's Syndrome, a fact which drove the depression-prone Paul further into increasingly erratic behaviour. Burdened with a large mortgage, the cost of Vassily's care in Jena (where Tatyana brought him for specialised treatment in 1922), and his own sense of self-doubt about being worthy of his position, Paul's emotional state spiralled persistently downwards, dragging the Ehrenfests' marriage with it. In the 1920s, Tatyana regularly returned to Russia, where she cast down the gauntlet in 1924 to the entire Russian educational system, arguing that the proponents of both the abstract and practical pedagogical approaches were wrong, and that a new approach that allowed practical intuitions to develop over time into abstract realisations was what was needed to revitalise the nation's mathematics programmes. Paul remained in Leiden, isolated in his room, despairing of his declining importance, carrying on a desultory affair with a woman ten years his junior, and watching the disturbing rise of anti-Semitism in Europe. His general sense of hopeless despair worsened until, in 1933, in the waiting room of the Amsterdam medical facility where Vassily went for treatment after his transfer from Jena, something snapped in Paul, and he shot and killed Vassily, followed by himself.

After the death of her son and husband, Tatyana returned to Leiden, where she remained until her death in 1964. The Soviet Union that had attracted her in the 1920s as a country devoted to real and substantive change in how societies function repelled her in the 1930s as it gave itself increasingly over to the orchestrated brutalities of Stalinism. These were difficult years, with the tragedy of Paul and Vassily's death compounded in 1939 by the loss of another son in an avalanche. In 1925, Tatyana had published a rigorous paper on thermodynamics that sought a clear distinction between the equilibrium state of a system and its irreversible progression over time, in many respects a far-sighted and revolutionary approach, which fell entirely in between the cracks of an academic community obsessed with relativity and quantum physics, and this relative disinterest in her thermodynamic ideas would persist through the 1930s. She expanded her 1925 paper into a full book treatment of thermodynamics, *Die Grundlagen der Thermodynamik*, which she had difficulty finding a publisher for. Sending the manuscript to Einstein in 1947, she received the reply that the book as it stood was too concerned with minutely nailing down its fundamental axioms and definitions, and that the main thread of the story was lost in all the detail. He declined to help find it a publisher, and did not reply to Tatyana's request for further advice.

Die Grundlagen der Thermodynamik was eventually published in 1956, but, as Einstein had predicted, her attempt to create a rigorous axiomatisation of thermodynamics failed to make many academic waves in spite of the several reigning paradoxes her system was able to resolve. A similar fate befell a 1958 paper on probabilistic motion for the *American Journal of Physics*. Her thoughts on education, and in particular about the early development of mathematical intuition as an important component of pedagogy, were more continually impactful, and in 1961 her essays on the subject were collected and published by Bruno Ernst. After a lifetime of tragedy and bad timing permeated by flecks of genius and rightful recognition, Tatyana Ehrenfest-Afanassjewa passed away in 1964 in Leiden. Her daughter, Tatiana Pavlovna Ehrenfest (1905–1984) became a mathematician in her own right, studying De Bruijn sequences and the BEST theorem.

FURTHER READING: For years and years, there was no single good source to go to for Tatyana Ehrenfest-Afanassjewa, and then suddenly, in 2021, *The Legacy of Tatjana Afanassjewa* was published by Springer, featuring articles about her life, her contributions to mathematical pedagogy, and her thermodynamics ideas. Since it is a Springer book it is, of course, absurdly expensive, coming in at over $80 for a *paperback* copy. The authors also made the strange choice to leave 'Ehrenfest' out of the title and to use a spelling of 'Tatjana' which is perhaps more representative of the right spelling, but is different from how you generally see her name spelt, so anybody looking for her most commonly given names of 'Tatiana Ehrenfest' or 'Tatyana Ehrenfest' on Amazon will simply not be able to find it, which only makes things more difficult for people trying to read more about her who haven't happened to buy this book pointing them towards it. In any case, if you have the money, it is pretty much the book to get. If you don't, I'd recommend getting the nice and inexpensive Dover edition of *Conceptual Foundations*, which features a reprint of the 1959 translation. Ehrenfest does the usual scientific widow thing that we saw in the last volume in the case of Margaret Huggins of writing herself out of the creation of the book in her introduction, but what follows is a beautiful tour through Boltzmann theory, clearly laid out and explained.

Chapter 20

Master of Matrices: Olga Taussky-Todd and How One of Mathematics' Coolest Objects Refound its Groove

In secondary school, matrices don't get anything near a fair shake. In case it has been a while, matrices were those rectangular grids of numbers that look something like this:

$$\begin{bmatrix} 5 & 2 & 3 \\ 1 & 4 & -2 \end{bmatrix}$$

In most secondary schools, when you are around 15, you learn the rules for adding them together, multiplying them, and for using them to represent systems of equations (the above matrix, for example, could represent the system $5x + 2y = 3$, $1x + 4y = -2$), and that is basically it. Throughout all of calculus, and then all of multi-variable calculus, they are nowhere to be seen, and it is not until the end of scientifically inclined students' first year in college that matrices reappear as the star players in a Linear Algebra course.

That perception of matrices as these odd objects that show up for two weeks mid-year to help out with systems of equations we don't want to deal with on our own, is emblematic of how, for many decades in the early twentieth century, the mathematical community itself considered the humble matrix – as an object to be used in the pursuance of some more noble task in some other, more important, field of applied mathematics.

That perception began to change throughout the 1930s and 1940s thanks to a small cadre of mathematicians who took this unloved stray, found the beauty in it, and began assiduously investigating its profound theoretical nooks to then share with the world. Few in this era championed the matrix and its mysteries so prolifically or profoundly as Olga Taussky-Todd (1906–1995), author of over 300 papers and for many decades the living nerve centre of the world's matrix community. Her knowledge of the matrix literature, proposal of new and intriguing problems, advocacy of the beauty and universality of matrix theory's under-appreciated results, and mentoring of two generations of researchers, all contributed to making the matrix, by the time of her death in 1995, the discipline-spanning powerhouse it remains today.

Olga Taussky was born in 1906 in Olmutz, at the time a city of the Austro-Hungarian Empire which is today the Czech city of Olomouc. She and her two sisters were encouraged by their mother in their various scientific pursuits such that the older sister became a professional chemist, the younger a pharmacist and clinical chemist and Olga, the middle child, one of the world's most revered mathematicians. Taussky, after a youth spent solving practical equations deriving from her father's work managing a vinegar factory, and earning money for the family as a private tutor after the early death of that father, entered the University of Vienna in 1925 to study chemistry.

These were the waning days of the University of Vienna's Golden Era – the philosopher and logical positivist Mortiz Schlick was lecturing there and holding meetings of his famous Vienna Circle (which Taussky attended and which also included the mathematician Hans Hahn and philosopher Rudolf Carnap) to discuss the ramifications of Ludwig Wittgenstein's works for the future of linguistic theory. Kurt Gödel was a student there at the time, and became a friend of Taussky, and with all of the exciting work being done in Vienna on the intersection between maths, logic and language, it was hardly surprising when Taussky switched her focus from chemistry to mathematics and in particular the world of number theory that had been introduced to her by the mathematician Philipp Furtwängler (who was, incidentally, the second cousin of the conductor Wilhelm Furtwängler, in case you're like me and that is the first thing you wondered).

In later interviews, Taussky always maintained that Number Theory was her first and true mathematical love, and during this time she dug deep into the world of class field theory that was emerging at that time thanks to the work of Furtwängler, Takagi, Hasse and others in tackling various problems and conjectures laid out by David Hilbert as to how a field of numbers K can be extended to a larger field L such that the prime ideals in the ring of integers associated with K can be factorised by the prime ideals of L's associated ring of integers. In less jargonistic terms, how do we expand a set of numbers so that a group of primes with a certain property in that set can be represented as a product of elements in the new, expanded set? Or, less jargonistic still, how do we expand our mathematical universe to unprime primes? For instance, if our set is the real numbers and the primes we are interested in are all of the primes p that have a remainder of 1 when you divide them by 4, then the set of complex numbers would be the expanded set you're looking for, because every such p can be represented as a product of two complex terms: $p = (a+bi)(a-bi)$. (Try some! It's fun! $13=(2+3i)(2-3i)$, $17 = (4+i)(4-i)$, $61= (6+5i)(6-5i)$ – by expanding your space, you just made unfactorable things factorable!)

Taussky got her PhD in 1930 for extending one of Furtwängler's results and for showing that further extensions could not be contained within one large universal rule, but would have to be treated and solved separately. Taussky spent the next two years at the University of Göttingen at the invitation of Richard Courant to work as an editor for the number theory sections of the upcoming collected edition of David Hilbert's works. Göttingen was another world capital of mathematics in the early 1930s, and it was here that Taussky's path crossed that of arguably the most famous woman mathematician of the early twentieth century, Emmy Noether. As mentioned in Chapter 14, Noether was a giant in the early development of abstract algebra who arranged for Taussky to run a course in class field theory to give the young mathematician a chance to gain experience in lecturing, and to give Göttingen a chance to hear what was going on over in Vienna.

By 1932, however, the political situation in Germany for a young woman of Jewish descent like Taussky was fast growing untenable, and so she sought out a position in England, winning a three-year fellowship at Girton College, the first of which she spent in the United States at Bryn Mawr, where Emmy Noether had relocated. Noether gave lectures every Tuesday at Princeton, which Taussky attended, and through which she met Albert Einstein and a number of other physics and mathematical luminaries who would play important roles in her life two decades later when she and her husband were professors at the California Institute of Technology.

With the early death of Emmy Noether in 1935, Taussky lost most of her reason for staying on at Bryn Mawr, and so returned to a Girton College where nobody seemed to share her particular interests. Collaboration is a key element in mathematical research, having people who share your interests and goals to bounce ideas off, and in the absence of that interaction it is difficult to make headway. Taussky left Girton for Westfield College in 1937, where she was given a crushing teaching load of NINE classes – made perhaps more bearable by the entry into her life of fellow mathematician John Todd, whom she would marry in 1938. It was a strong marriage of minds engaged in similar pursuits that would last nearly six decades, until Taussky's death in 1995.

However, 1938 was an unfortunate year to begin a new married life, as the outbreak of war in 1939 brought with it instabilities in work, housing and food availability that the young couple struggled against, relocating themselves eighteen times over the course of the war while both awaited positions that would allow them to use their mathematical talents for the war effort. In 1943, Taussky was given a position at the Ministry of Aircraft Production investigating the phenomenon of wing flutter, part of which analysis came down to the conditions of stability of a matrix. She had already been working on questions dealing with matrix theory, but this application of a theoretical aspect of a matrix to an applied outcome put her on the scent of how beautifully matrix theory reached into other parts of not only mathematics, but the wider world.

At war's end, John and Olga received invitations to work for the National Bureau of Standards in the United States. They worked at NBS for ten years, from 1947 to 1957, where part of Taussky's job was to read every paper on matrix theory submitted to NBS so that, by the end of her time there, she had a familiarity with matrix literature that routinely astounded other mathematicians. At NBS her title was 'consultant in mathematics', a vaguely defined position that meant essentially doing whatever odd mathematical job the NBS had lying about, from indexing the papers written by its researchers, to proposing novel problems to solve, to assisting the bureau's researchers, to answering letters from the public.

Somehow, in spite of this load of vast and ill-defined responsibilities, Taussky had time for her own research, and began publishing the papers that would rekindle the broader mathematical community's interest in matrix theory's under-appreciated back alleys. Her 1949 paper 'A Recurring Theorem in Determinants' revived the Diagonal Dominance Theorem, which had been first published in 1881 but which had, by 1949, lain neglected for the better part of a decade and a half. That theorem has to do with what happens when a square matrix (one with equal numbers of rows and columns) has a main diagonal (the diagonal running from the upper left to the lower right) with very large terms in it. In particular, what happens when each diagonal entry is greater than the sum of all the other numbers in its row, as in the following matrix?

$$\begin{bmatrix} 8 & 4 & 1 \\ 4 & 9 & 3 \\ 3 & 1 & 6 \end{bmatrix}$$

Well, something clever happens, as it turns out. For a matrix with such a dominant diagonal, a quantity called the determinant will never equal 0, which means that the matrix has an inverse, that a solution to the system represented by the matrix exists, and any number of other consequences that you usually have to do some finger-numbing matrix massaging to find out, but that this theorem lets you determine at a glance.

(By the by, this 1949 paper also brought back to light Semyon Gershgorin's neglected 1931 theorem about where in the complex plane the eigenvalues of a matrix can be found, which Taussky had found useful in her wartime matrix stability work. Gershgorin's method would prove to be a powerful way to approximate the eigenvalues of a matrix for a pre-calculator age that needed every calculation tool it could get.)

In 1951 she and T.S. Motzkin presented 'Pairs of Matrices with Property L' to the American Mathematical Society, which contained a beautiful result relating to matrix commutativity. In everyday life, we are used to the fact that multiplication is commutative, i.e. that if you do 2 x 3 you'll get the same answer as if you did 3 x 2. Order doesn't matter. With matrices, however, this is not the case. There is no guarantee that if you multiply matrices A and B in the order AB you'll get the same result as if you multiplied them in the order BA. What the Taussky-Motzkin paper established was a condition that, if satisfied, guaranteed that $AB=BA$ (if you're curious, that condition is that the matrix $aA + bB$ is diagonalisable for any values of a and b).

In 1957, John and Olga were scooped up as a pair by the California Institute of Technology (Caltech) which had not, up to that point, had a woman on the faculty of its mathematics department. She was, however, only a 'research associate', which was a step down from her tenured position at NBS, and it would not be until 1971 that she received a full professorship and thereby became the first woman at Caltech in any department to hold that title. If her initial title was underwhelming, however, the work was enlivening. She loved teaching advanced matrix theory to graduate students and working closely with her PhD students, over a dozen of whom earned their PhD under her encouraging guidance. She also continued to propose problems that tantalised the mathematical world and to demonstrate the power of neglected matrix theories, as in her 1961 resurrection of Lyapunov's Theorem, first hypothesised in 1892 in the context of the stability of differential equations that model dynamic systems, and which, like Gershgorin's Theorem, makes important claims about where the eigenvalues of a given matrix can be found.

Olga Taussky-Todd wore every hat in the mathematician's cupboard over the course of her long career: applied researcher; abstract theoretician;

lecturer; mentor; archivist; problem poser; historian; and public relations official. Most professors find performing two of those activities well to be the limit of what their time and energy can manage, but Taussky-Todd, to all accounts, was a gracious and kind virtuoso who leapt between those roles with poise and infectious enthusiasm. To no one's surprise, when she hit Caltech's mandatory retirement age of 70 and became a professor emeritus, she continued, against the university's stated policy, to supervise doctoral students, and to nobody's regret, the university decided to look the other way and let her continue to inspire another generation of students, regulations be damned.

Olga Taussky-Todd died in her sleep on 7 October 1995.

FURTHER READING: Edith H. Luchins wrote both the chapter on Taussky-Todd in *Women of Mathematics: A Biobibliographic Sourcebook* (1987) and the American Mathematical Society's obituary of Taussky-Todd in the *Notices of the AMS* (1995, Volume 43, Number 8). The latter you can get online, but the former has more details of her life and background. For her mathematical impact, I quite like Hans Schneider's tribute to her place in matrix theory, 'Olga Taussky-Todd's Influence on Matrix Theory and Matrix Theorists' written in 1977, and available online. There is an *In Memoriam* volume to Olga Taussky-Todd put out by International Press, but as a source detailing her life and work it is probably not what you're hoping for.

Chapter 21

Varieties: The Life and Mathematics of Hanna Neumann

Of all the realms of mathematics, there are few where more people feel more at home than in the safe harbours of algebra. From the age of 8 or 9, we are made familiar with the classic moves involved in Solving For X, how $a(b+c) = ab + ac$, how if $a = c$ then $a-c = 0$, and a fleet of other manipulations besides that become so second nature to us that we rarely stop and think about just what we are doing. What sort of a thing is it we are engaging in when we employ the laws of algebra, and might the moves that we have got used to be nothing more than very restricted cases of a much bigger world?

These questions began intriguing mathematicians in the late nineteenth century, leading to the explosion of the field of Universal Algebra in the 1930s and 1940s, which sought to answer the question of what makes algebra algebra, and how we might construct things that follow the rules we most associate with the algebra we've learned since youth, but which present us with often strange new mathematical vistas. Some of the most interesting results of universal algebraic thinking came when mathematicians began playing with the properties possessed by different categories of algebras, including the field of 'Varieties of Groups' pioneered by the mathematical power couple of Hanna and Bernhard H. Neumann.

Hanna's mathematical journey would take her all over the world, first in flight and later in triumph. Born Hanna von Caemmerer on 12 February 1914 to a family of long-standing Prussian military background, she lost her tradition-flouting historian father early in the First World War, and her Huguenot-descended mother had to make do on a slim war pension that required she and her two siblings to contribute to the household as best they could. Hanna's academic gifts were such that, at the age of 13, she was able to earn money coaching other children. From 1922 to 1932 she attended the Augusta-Victoria Schule, where she took fifteen subjects and, on her final examinations for college, scored highly on all areas except music.

Hanna began her first year at the University of Berlin in 1932, where the foundational topologist Georg Feigl and function theorist Ludwig

Bieberbach were among the mathematical luminaries offering courses to undergraduates. Hanna allowed herself to experience the full breadth of the university's offerings, taking courses from Gestalt theory psychological pioneer, Wolfgang Kohler, Nobel prize-winning physicist Walther Nernst, and Germany's most famous academic lawyer, Martin Wolff. She threw herself into the coffee break academic discussion culture of the university, and soon became good friends with doctoral student Bernhard H. Neumann.

Their romance, for soon it became such, would determine the course of their lives, for with the ascension of the Nazis to power in 1933, Bernhard, of Jewish descent, knew his days in the German university system were numbered, and emigrated to England to continue his studies at Cambridge. He and Hanna became engaged secretly to avoid reprisals against the members of Bernhard's family still in Germany, though Hanna soon caught the official attention of the Nazis anyway by her work with a group of fellow students who sought to keep non-enrolled Nazi rabble rousers out of classes taught by Jewish professors. Hanna continued her mathematical studies, but in her fourth year she was faced with the decision of applying for her doctorate, which is what she wanted to do, or taking the Staatsexamen, which was a broader degree for those desiring to go into public service and the teaching profession. Unfortunately, she knew that if she opted for the doctorate, a pro-Nazi member of the administration would be sitting on her examination panel, and would question her about her 'political knowledge', which would serve as a thinly veiled evaluation of her Nazi political purity.

Hanna elected to take the Staatsexamen instead, which she completed in 1936, and then accepted a position as a research student to Helmut Hasse at the University of Göttingen, which was, by 1937, a shadow of its pre-Nazi self. But with the Nazi territorial expansions of 1938, Hanna realised that it would no longer be possible to simply wait out the regime, and she broke off her course of study early to go to England. Here, she and Bernhard could be together, though lingering fear for Bernhard's family compelled them to marry secretly in 1938. Hanna continued her work (mostly, thanks to the housing problems created by the Second World War, on a card table in a caravan parked near a haystack on a market gardener's farm) about the 'subgroup structure of free products of groups with an amalgamated subgroup', under the light supervision of Olga Taussky-Todd.

Let's break this down a bit.

The area of mathematics that Hanna Neumann worked in is one that, taken piece by piece, is something anybody could probably understand, but it is so enwrapped in specialised vocabulary that few make the attempt. Neumann's most enduring work was in the area of 'Varieties of Groups', about which she wrote a book in 1967 that is still a widely referenced classic. To understand the field she was working in, you need to know about mathematical varieties, and to understand *those*, we need to go back to our opening question – just what is an algebra, anyway?

Basically, when we talk in the most general possible terms about an algebra, what we're talking about is a set of numbers or elements, with a set of operations and a set of defining equations. The operations are mathematical actions that combine elements from the number set and produce an element from that same set. They are classified on the basis of how many elements from the number set they need to function. Addition is a 'binary' operation because, to add two numbers together you need ... *two* numbers! Inversion is a 'unary' operation because all I need in order to give you the inverse of a number is the *one* number you want the inverse of. There are even 'nullary' operations that you don't need any input for because the output number never changes – for example, if you want to know what number, added to any number, gives that number back again, I can tell you the answer is '0' without you providing me with an input value. That is just how zero works, regardless of what it is being combined with.

So much for the operations, what about the defining equations (or axioms as they're sometimes known)? Well, these are just the special properties that we want this number set, with whatever operations we've loaded on top of it, to possess. The most common universal algebra is the Group, which has three operations – a binary one (like addition or multiplication), a unary one (inverse finding), and a nullary one (the existence of an identity element that always results in the return of the initial input) – and three axioms, which you have used since elementary school:

> Associativity: Whatever your operation is (I'm going to call it *), it has to be true that $(a*b)*c = a*(b*c)$. You're allowed to do the operations in whatever order you want.

> The Identity Axiom: If 'e' is your identity element in the set, then $x*e=x=e*x$.

> The Inverse Axiom: If $\sim x$ is the inverse of element x, then $x*\sim x = e = \sim x*x$.

Most of us have lived our lives within the thin slice of the algebraic universe given by the Group defined by the integers with the binary operation of addition, and another algebraic structure called a Ring, which allows you to have two binary operations, like the real numbers under multiplication and addition. But there are so many more out there waiting to be played around with, and once you realise that it is natural to ask the question, if there are multiple possible different things out there called algebras, how do we start grouping them together, and what properties do those collections of algebras possess?

Enter the 'variety'. A variety is just a set of algebraic structures, all of which share a given 'signature' (set of binary, unary and nullary operations), and which have the same rules for determining when two things are 'equal' to each other, or alternately, what combinations of elements can be counted on to equal 1.

Hanna Neumann's work explored the properties of varieties, and in addition to proving some important theorems (such as the proof that all finite free products of finitely generated Hopf-type groups are themselves Hopf-type) she proposed a number of stimulating questions and challenges that motivated the field throughout the late-mid-twentieth century. She received her PhD in 1944, and was given a DSc by Oxford in 1955 in recognition of her group theoretic publications, and somewhere in the interim from 1939 to 1951 found the time to have five children.

From 1946 to 1958, she lectured at the University College of Hull, where she was instrumental in injecting the joys and challenges of pure mathematics to a curriculum that focused on rote applied problem solving. She brought that same spirit of horizon-opening enthusiasm to her next position, overseeing the pure mathematics curriculum at the University of Manchester, and introducing, thereby, a generation of young British mathematicians to the most advanced methods and exciting prospects from the European abstract tradition, while she herself expanded her studies of the properties of varieties that are created by groups, which are themselves built from the combination of a finite set of elements.

Finally, in 1963, she and Bernhard were offered positions at the Australian National University, which they accepted, and where they would remain until Hanna's death in 1971. In Australia, she devoted herself not only to her research and department but to the wider, and perhaps more vital, question of secondary school curriculum restructuring, and the perpetual problem of preparing a nation's teachers, all of vastly different ages and experiences, how to cope with the new standards. She was elected in 1966 as vice president of the Australian Association of Mathematics Teachers upon the founding of that group, and used her intercontinental experience of different mathematical teaching systems to help guide the creation of a series of pamphlets to help teachers grasp the new mathematical topics being introduced to the Australian curriculum. In recognition of both her contributions to pure mathematics and to Australian education, she was elected in 1969 as a fellow of the Australian Academy of Science.

The last years of her life were spent travelling the world, introducing her ideas (including her recent Hopf proof) to audiences in Germany, Canada, the United States, and France, where her natural ear for languages stood her in good stead. In demand as a speaker, educational reformer, department overseer and fundamental researcher, with a growing and successful family, she had every reason to expect several more decades of full life when, on tour in Ottawa, she checked into a hospital the evening of 12 November 1971, reporting that she was feeling slightly ill. That night, she slipped into a coma, and died on 14 November.

FURTHER READING: Neumann's classic *Varieties of Groups* (Springer-Verlag 1967) is not too hard to find, but is not a particularly inviting introductory text to the topic. To get your foot in the door, I'd recommend Dummit & Foote's classic *Abstract Algebra* text, which introduces what you need to know about groups, rings, fields and various morphisms, from essentially the ground up. For her life, your best source is probably the Australian Academy of Science memorial to her, which is available online.

Chapter 22

Trajectories: Katherine Johnson's Orbital Mathematics

Before NASA, there was NACA, an oddball collection of aeronautics nerds using black box data and wind tunnel analysis to figure out as much as they could about the science of flight. Calculations, done almost entirely by hand, were the coursing lifeblood of the organisation. Those calculations were handled by a small army of women who were 'checked out' and 'returned' to the mathematical pool as needed by the male scientists. And in a room separate even from *those* women was the place where the African American calculators were kept, segregated because of their race, and given brute force computational tasks because of their gender.

All the externals spoke against sustained success for any non-white female mathematician and yet, from within that segregated room there came one person, Katherine Coleman Goble Johnson (1918–2020), who refused to be domineered by tradition and who, as a result, over thirty-three years at NACA and NASA, made fundamental contributions to the Mercury and Apollo missions, the space shuttle, the mathematics of space flight, astronaut emergency navigation systems, satellite tracking techniques, *and* plans for a future Mars mission.

She was a maestro of trajectories, NASA's go-to mathematician for developing the equations of the country's first ventures into space and, as could be expected, she showed genius from the start. Her parents cared about education more than anything, and her father was willing to work in one city while the rest of the family lived in another just to ensure that she had access to a high school. He only got to see them once a month when school was in session, but thought the sacrifice was worth it, especially as Katherine continually leapt with great bounds over and beyond her classmates.

She entered school at the second grade, bypassing kindergarten and first grade entirely and then her teachers found her still so advanced that they skipped her over fifth grade as well, with the result that she entered high school at the age of 10. She chewed through the standard mathematics courses

103

handily and so her teachers developed a college-level course, which had her as its only student. She attended West Virginia State, a historically African American school, and benefited from the intense and personal mentorship of her professors there. Professor John Matthews, a fluent speaker of seven languages and head of the Romance Languages department, inspired her to study French and English, while Professor James Evans, who took her into his family as an adopted daughter of sorts, pushed her to continue expanding her substantial mathematical skills, with the result that she double majored in French *and* mathematics, getting her degrees at the age of 19.

Had she been a male, she would have been swooped up into the field of research mathematics that Evans encouraged her to pursue, but as an African American female in the late 1930s, teaching was more or less her only option, taking what jobs she could as they arose to keep herself out of the grinding maw of the Great Depression. For fifteen years she carried on like this. For two years she attended graduate school but had to leave before obtaining her degree when her husband, who she'd married just after graduating West Virginia State, started succumbing to brain cancer. She dropped her studies to find more work as a teacher to provide money for his care and for the raising of their three daughters.

For most women struggling to feed their family during the Depression, that, or something still more tragic, would have been the end of it, but fortune granted Johnson a break. She heard about jobs for women with mathematical abilities at NACA. Not glorious jobs – primarily grinding out routine calculations that the aeronautic researchers and engineers couldn't be bothered with, but real and steady work at the heart of a new frontier. Johnson joined the pool of calculatorial ladies in 1953, and was assigned her spot in the 'colored room' to wait until she was needed.

Classified as a 'subprofessional' on account of her gender, she used slide rules and instinct to churn her way through masses of wind tunnel data. Johnson was so deft mathematically, however, that she was soon snatched up by the Flight Research Division, who decided to keep her rather than returning her to the lady calculator closet. She argued for, and won, a place at division meetings, a thing unheard of at the institution, but one that was generally accepted because her deep knowledge of the subject was too useful to have at hand to refuse.

When, in 1958, NACA became NASA and the entire organisation had to shift its focus to compete with the Soviet Union in space, they found themselves in the embarrassing situation of having no basic text laying out the mathematics of space flight, and Johnson was put on a secret team to create it. That work turned out to be one of NASA's core manuals during the institution's Golden Age. Johnson was then let loose on trajectory calculations. When a new mission was being planned, a representative would come to Johnson and ask for the trajectory equations that would result in the desired touchdown location and time, and she developed them, creating the trajectories that lifted Alan Shepherd and John Glenn into space, and which handled the immense task of getting the Apollo programme to the moon, the lander to the surface, and the crew back to Earth.

In an interview with Wini Warren, she related:

> Everything was so new. The whole idea of going into space was new and daring. There were no textbooks, so we had to write them. We wrote the first textbook by hand, starting from scratch. People would call us and ask, 'What makes you think this or that is possible?' and we would try to tell them. We created the equations needed to track a vehicle in space. I was lucky that I was working with the division that worked out all the original trajectories, because I guess that's what I'm remembered for.

She worked days and nights, pulling sixteen-hour shifts with her colleague Al Hammer in order to develop fail-safe routines that astronauts could use to navigate by the stars if their link with the ground was severed or the computers failed, and in 1970, when Apollo 13's systems were damaged and it was feared the crew wouldn't make it back, it was Hammer and Johnson who were summoned to track the flight and advise the white-knuckled administration.

While working on these problems for NASA, Johnson wrote her first scientific paper in 1960, 'Determination of Azimuth Angle at Burnout for Placing a Satellite over a Selected Earth Position' which, upon its publication in 1960 represented the first time NASA had ever allowed a woman to attach her name to the work she did in a public document. It was instrumental in the planning of the Mercury missions, and would be followed by twenty more technical papers over her thirty-three-year career. After Apollo, she worked on finding better procedures for object tracking, work which required sifting through tracking station data from all over the world and which helped make possible our satellite-clustered sky of today.

She worked on the space shuttle, the projected Mars programme, and on methods of using satellites to map the mineral resources of the planet. Her projects read like a Greatest Hits album of humanity's efforts in space, from its slide-rule and wind-tunnel origins to its dreams of interplanetary exploration and satellite proliferation. She started in a segregated office trying to earn enough to keep her husband well and her family thriving. She ended by mathematically mapping the expansion of humanity from its modest home planet. 'I had a wonderful, wonderful career at NASA. I don't imagine everyone can say that, but it's true in my case. I have always loved the idea of going into space – I still do.'

Katherine Johnson passed away in a retirement home on 24 February 2020, at the age of 101.

FURTHER READING: The publication of *Hidden Figures* by Margot Lee Shetterly in 2016, and subsequent 2019 film, lifted Katherine Johnson to a level of well-deserved scientific superstardom that paved the way for the publication of her autobiography, *My Remarkable Journey: A Memoir* in 2021. Before 2016, your best source for her life and work was contained in Wini Warren's *Black Women Scientists in the United States* (1999), which is a key text for anybody interested in the history of gender and race in the sciences and features easily a hundred biographies of success torn from the mouth of systemic discrimination.

Chapter 23

Julia Robinson and the Cracking of Hilbert's Tenth Problem

For mathematicians, the only thing more exciting than proving a theorem is proving that it can never be proven. These anti-proofs, if you will, stand firmly against all future progress of humanity and state, 'No matter how clever you become, what new branches of thought you invent, you'll never be able to do this. Sorry about it.' The most famous of these is Kurt Gödel's 1931 Incompleteness Theorem, but just behind it in the annals of mathematics is the 1970 proof by Yuri Matiyasevich that Hilbert's famous Tenth Problem will never be solved, a proof that might never have happened without the almost other-worldly mathematical insight of Julia Robinson (1919–1985).

Robinson's childhood story is perhaps a familiar one. Take a look at any of her class photos, and she sticks out instantly – round owl glasses perched just beneath straight, unstyled hair and above an oversized mouth, all kept in frame by a thick stalk of a neck. She was, from the first, somebody who didn't care about what other people thought of her, and never really changed, too wrapped up in her inner world to pay much attention to such slight things as outward appearance.

She was also subject to constant ill health. At the age of 9, she caught scarlet fever, and hardly had that subsided when she was hit with a round of rheumatic fever, which permanently damaged her heart, followed by chorea, which periodically racked her body with spasms.

After years in and out of school on account of her health, she caught up on her schoolwork and went to San Diego High School, where her talent for problem solving kept her in maths and physics classes long after most girls dropped them for other avenues of study. She was routinely the only girl in those classes, and the best student as well. Graduating, she received a new slide rule (named 'Slippy') from her parents and started her college career at San Diego State University, where the mathematics department wasn't much of much. They offered two upper division maths courses a semester, and that was it.

She might have stayed in that programme had it not been for an encounter with E.T. Bell's magnificent book, *Men of Mathematics*. In it, Robinson saw at last the full grandeur of mathematics, and by implication, how very much she was missing in SDSU's limiting environment. She made up her mind to go to Berkeley, where the mathematics department was just being shaped into the powerhouse it would become by Griffith C. Evans.

There, she had access to the resources she needed at last to develop into a full mathematician. She took courses in number theory, the discipline that would define her for the rest of her life, from Raphael M. Robinson, whom she married soon thereafter. Her thesis advisor was Alfred Tarski, who in 1939 had made an important addition to Gödel's Incompleteness Theorem, which makes this as good a time as any to get into just what that was all about.

Gödel's theorem of 1931 states that, if you have a set of statements which contains only natural numbers, the number zero, the successor, addition, and multiplication operations, the logical operations 'and', 'or', and 'not', and the expressions 'for all' and 'there exist', then that set must contain statements that cannot be proved or disproved using the Peano axioms (the axioms for the natural numbers like that $a=a$, or that $a=b$ implies that $b=a$).

Natural number theory, then, contains statements that cannot be proven using a general algorithm, a result which leaked into philosophy and caused a small cadre of hasty academics desperate for copy to assert that room for God had been found at last in maths. Alfred Tarski was curious about whether the Incompleteness Theorem applied to other sets of numbers, and set his sights on the Reals, which are a much denser set of numbers than the Naturals, and offer a much richer set of analytic tools and theorems. In 1939, he discovered that, in fact, the Real numbers don't suffer from incompleteness as the Naturals do, and that you *can* construct an algorithm for deciding the truth of any real-number-based statement.

Real number theory is, in the terms of number theory, 'decidable' – you can decide whether a random real number statement is true or not using a general algorithm. Natural number theory is not. What was left to decide was the classic middle ground between integers and Reals – the Rationals.

Remember that Rational numbers are those that can be represented as a ratio of integers. They contain the integers and natural numbers (5 can be represented as 5/1), and are contained in the Reals, but don't contain those Reals which cannot be represented as fractions – numbers like pi, or the square root of 3. The question was, are the numbers you lose when you move from the Reals to the Rationals enough to make the Rationals as undecidable as the Natural numbers?

This was the problem put to Robinson, who characteristically decided to solve it by referencing an obscure set of theorems which, to all appearances, had nothing to do with the decidability of the Rationals. When her colleagues described her, this was the aspect of her work that all of them hit upon at some point – the 'How did you possibly think to look *there*?' aspect of her mathematical mind. Using a set of obscure theorems, and employing her own innate brilliance to craft from them a function that would break the decidability of the Rationals if it could be shown to have the properties she thought it did, she proved at last in her PhD thesis that the Rationals were undecidable, brethren to the Naturals.

This work turned out to be excellent training for Hilbert's Tenth Problem, the ultimate unsolvability of which she spent two decades attempting to establish, only to have the final stroke of the anti-proof pulled off at last by a 22-year-old Russian mathematician in 1970. In 1900, the mathematician David Hilbert had proposed a list of twenty-three problems for the new century to grapple with. Solving any one of these is a relatively sure way to mathematical immortality, and today only three remain unsolved. Hilbert's Tenth Problem (hereafter H10) was to find a general algorithm that would determine if any Diophantine equation with integer coefficients was solvable.

Diophantine equations, as we remember from our time with Hypatia, are just polynomial equations in several variables for which we only accept integer solutions. For example, $x^2 + y^2 = z^2$ is a Diophantine equation in three variables with (3,4,5) as one allowable solution. Hilbert challenges us to develop a general method which tells us whether or not a random Diophantine equation has an integer solution or not. It doesn't ask us to find the solution, it just asks us to determine whether or not a solution exists.

Surely, that is a reasonable request, isn't it? Just one little general method that says 'yes' or 'no' each time it is presented with a new equation – is that too much to ask?

Spoiler alert: Yes.

It turns out that such a method does not exist, cannot exist, will never exist, and the foundation for proving this was laid down by Julia Robinson. Robinson realised that, to evade being decidable by an algorithm, she would have to construct a Diophantine equation whose roots showed exponential growth. She would need to show, in effect, that exponentiation (raising a number to a variable power) was Diophantine, which meant hunting for an equation A with three parameters a, b and c and a finite number of unknowns whereby $A(a, b, c, x1, x2, ..., xm) = 0$ has a solution in $(x1, ..., xm)$ if and only if $a=b^c$.

If you found that function, it would let you change an exponential Diophantine equation into a regular Diophantine equation, which would mean that exponentials, with their much nastier qualities, could be brought to bear to break any method that might claim to solve H10. In 1950, she simplified the search through another flash of insight that, to get to her H10-breaking function A, all one really needed was a function B with two parameters, a and b, which has the properties (1) that $a < b^b$, and (2) for any integer value of k, there exists an a and b such that $a > b^k$.

That such a Diophantine relation of exponential growth exists became 'The JR Hypothesis' and it was recognised that, should it be proven true, should a Diophantine equation be found that fit those properties, it would admit exponential growth into Diophantine equations, and wreck utterly any hopes of finding an algorithm which determines the solvability of all Diophantines. Robinson proposed the JR Hypothesis in 1950 and had to wait two decades until Yuri Matiyasevich discovered the equation and resultant set which fulfilled the JR Hypothesis and thus rendered H10 definitively unsolvable.

(If you're curious, he found the solution, of all places, in the Fibonacci sequence – $a_{n+2} = a_{n+1} + a_n$ or 0, 1, 1, 2, 3, 5, 8, 13, 21 … If you define b as the ath even term in the Fibonacci sequence, the resulting set of a and b values matches Robinson's criteria and grows exponentially as $((3+\text{sqrt}(5))/2)^n$. Since the Fibonacci sequence contains all the numbers which solve the Diophantine equation $x^2 - xy - y^2 = +/- 1$, this is a Diophantine set which also happens to display exponential properties.)

Matiyasevich became a mathematics superstar at the age of 22 for his discovery and, in a tale which isn't told enough in the history of science, went out of his way to always include Robinson in any mention of the cracking of H10. Their correspondence and occasional joint work represents a relationship of total respect and consideration that serves as a model for young scientists recognising the work that went before them, and for veteran scientists gracefully acknowledging the gifts of a rising generation.

Formal honours followed in rapid succession, as Robinson became the first woman elected to the National Academy of Sciences in 1976, and the first female president of the American Mathematical Society in 1982. But she'll always be remembered as the person with the astounding ability to concoct math-breaking functions from the arcane theories of far-flung realms of mathematics, and her own inexplicable numerical genius.

FURTHER READING: From one perspective, Number Theory is not an easy branch of mathematics to jump into the middle of. Unlike analysis or hyperbolic geometry, whose basic ideas one is exposed to during a typical high school career, number theory is very much its own thing, and hardly touched in general maths classes beyond the classification of numbers as integers, Rationals, or Reals. From another, it is the ideal entry point for a person who has forgotten all their secondary school maths, but wants to return to the magic of mathematics, as all it involves at first glance are numbers that we are all pretty familiar with. For me, the YouTube channel *Numberphile* is a great entry point into the weird and wonderful world of thinking about numbers, and if that gets you hooked, you can head to Kuldeep Singh's *Number Theory: Step by Step*. Meanwhile, for Robinson herself, *Julia: A Life in Mathematics*, by her sister Constance Reid, is an interesting mixture of personal reflections, numerous rare photos, and a set of excellent essays by those who worked with Robinson, explaining her various discoveries with perhaps more rigour and detail than the average reader might want, but which real Robinson enthusiasts will definitely appreciate. If number theory is already your bag, Matiyasevich wrote an entire book, *Hilbert's Tenth Problem*, which delves deep into the history and ultimate anti-solution of the problem.

Chapter 24

Bringing the Computers of Earth to the Problems of Space: Evelyn Boyd Granville and the Early Days of IBM at NASA

The IBM 650 was a marvellous beast. The world's first mass-produced (and first profitable) computer, it was the mainstay machine of the 1950s, its magnetic drum capable of storing up to 4,000 words in its memory as it rotated at a rate of 12,500 revolutions per minute, allowing what was for the time a blazing capacity for computational speed. Marvellous as it was, it required a distinct and rare group of core competencies to operate, and the maestros of the device were in high demand by both government and private industry, which fell over themselves to attract people with the right combination of abstract numerical analytic abilities, practical computational rigour, and raw engineering instincts to make the 650 hum.

One of the great luminaries from this exciting era, when the world's first mass computer met humanity's first attempts to find its way into space, was Evelyn Boyd Granville (b. 1924). The second black American ever to earn a PhD in mathematics, she harnessed her knowledge of numerical analysis to her pioneering interest in the application of computers to scientific problems to garner for herself a succession of positions at IBM, NASA and the NAA, working on her era's biggest computational puzzles.

Born on 1 May 1924, the dizzying heights of Evelyn Boyd's future academic success were anything but assured. Her birth town of Washington DC was segregated, her father left the family while Boyd was a child, and the Great Depression struck her family when she was but 5 years old. Her single mother, Julia Walker Boyd, was left to raise Boyd and her sister, during a time of economic destitution, in a city of institutionalised racial discrimination, on the income she could manage from her work as first a maid, and then later as a stamp and currency examiner for the Bureau of Engraving and Printing.

Arrayed against these lingering adversarial forces, however, were a number of supports that proved crucial for Boyd's ultimate success. First, her aunt, Louise Walker, who was a graduate of the Miner Normal Teachers' College, was devoted to the cause of education as the black population's best way to lift themselves out of their economic hardship, and was willing to put her own savings on the line to ensure that the Boyd sisters had an education that would nurture their gifts. Second, though Washington DC was segregated in many of its facets, it also happened to boast one of the nation's elite academic institutions for black children: Paul Laurence Dunbar High School.

Founded in 1870 as M Street High School, it was the first public high school in the United States to serve the black population. In 1916, the school moved, and changed its name to Paul Laurence Dunbar High School in honour of the poet who had passed away in 1906. Whereas other high schools in the city were categorised as vocational in focus, Dunbar was designated as an academic achievement school, and attracted a crop of highly trained, deeply competent teachers, some of whom possessed doctorates in their fields, devoted to the principle of black educational excellence. As a result, over the subsequent years, Dunbar alumni included a robust roster of some of the nation's most famous and influential trailblazers, from Senator Edward Brooke, to the discoverer of blood plasma, Charles R. Drew, to John Aubrey Davis, who played a key role in the *Brown v. Board of Education* Supreme Court case that ended school segregation.

As educational opportunities for a brilliant young black woman went, this was the jackpot, and its assembly of successful, enthusiastic teachers served as constant inspiration for young Boyd, who wanted nothing more than to become like one of them. Her mathematical abilities were noted by teachers Ulysses Basset (a graduate of Yale University) and Mary Cromwell (a graduate of the University of Pennsylvania), who both encouraged her to apply to top-level colleges. Cromwell in particular suggested that she apply to Smith College and Mount Holyoke, both of which accepted her, but neither of which offered any type of scholarship. Fortunately, Aunt Louise offered up $500 from her savings to help Boyd meet the expenses of her first year at her chosen school, Smith, and she supplemented this money by taking summer work at the National Bureau of Standards, earning scholarships from Smith, and economising by living in co-operative housing for her last three years at the college.

While at Smith, she studied mathematics, theoretical physics and astronomy, the latter of which nearly tempted her to switch majors. Ultimately, the prospect of sitting in isolation for long observational stretches overcame her enthusiasm for the mind-expanding scope of a career in astronomy, and she settled in to a future as a mathematician. Boyd graduated *summa cum laude* in 1945 with honours in mathematics, and was accepted to both the University of Michigan and Yale for graduate studies. Boyd chose Yale, because it offered additional scholarships that Michigan did not, and worked there under Einar Hille, who was president of the American Mathematical Society from 1947–48, and whose name lives on in functional analysis textbooks in the form of the Hille-Yosida Theorem.

Boyd's research under Hille concentrated on Laguerre series, which are sums of the polynomials that solve the differential equation $xy'' + (1-x)y' + y = 0$, and which have importance in the realm of quantum mechanics. She received her PhD for this work in 1949, becoming only the second black woman in the United States to receive a Mathematics PhD (Euphemia Hayes was the first, in 1943, and Marjorie Lee Browne would be the third, also in 1949, while the fourth, Argelia Velez-Rodriguez, would not occur until 1960). Though she found research interesting, she was also profoundly pulled by the personal fulfilment of teaching, and spent two years as Associate Professor of Mathematics at Fisk University in Nashville, Tennessee, where she taught future PhD students Vivienne Malone-Mayes and Etta Zuber Falconer.

In 1952, she took the fateful step of accepting a position at the National Bureau of Standards in Washington DC. Here, she worked on mathematical analysis of problems related to the creation of missile fuses; more importantly, she met a group of mathematicians who were working at the NBS as computer programmers, using the power of the new machines to analyse problems arising from scientific research. She found the prospect fascinating, and when a chance arose in 1955 to work for computer industry leader IBM, she grasped it.

Joining in 1956, Boyd was just in time for the computing revolution created by the IBM 650, which was introduced in 1954 and would continue production until 1962. She learned how to work with SOAP, which was the assembler that attempted to semi-automate the optimisation of instruction addresses on the 650's magnetic drum. As mentioned, the drum rotated at 12,500 revolutions per minute, meaning that one revolution took about 4.8 milliseconds to complete. Part of the trick of writing code for the 650, then, meant placing instructions in memory so that, when one was completed, the part of the drum that contained the next instruction was in place to be read right away. If you randomly placed instruction addresses on the drum, you stood to vastly slow down computation speed, by having instructions finish, and then having to wait up to 4.8 milliseconds for the drum to rotate around to the location of the next instruction. SOAP helped this process to a degree by automating the placement of instructions so that, when one completed, the drum was exactly where it needed to be for the next instruction to be read.

Boyd learned how SOAP worked, and learned how to improve upon its automated assignments, while creating programs for the 650, and ultimately becoming a numerical analysis consultant for the Service Bureau Corporation, which was a subsidiary of IBM. This was the time when NASA realised the vast potential of computers to aid in trajectory computations, and awarded IBM a contract to essentially run the Vanguard Computing Center in Washington DC. Boyd saw the chance to apply her experience with computers and expertise in mathematical analysis to problems related to space travel, and volunteered to join the IBM group that would be overseeing Vanguard, working on the programs that calculated trajectories and mission models for Project Vanguard and Project Mercury. Her marriage in 1960 caused her to leave IBM and take up a job in Los Angeles for the Computation and Data Reduction Center of Space Technology Laboratories, where she worked on refining computational methods for determining orbits.

As the space race grew in intensity, so did NASA's need to farm out work to private industry, and in 1962 Boyd joined the North American Aviation Company, which had just received a NASA contract for design work related to the upcoming Apollo missions. Boyd worked as a specialist there, focusing on problems of celestial mechanics, orbital computations, digital computation methods, and numerical analysis – work she carried out until 1963 when she rejoined IBM to continue her research in using numerical analysis to improve trajectory computation. This work continued until 1967, when IBM began cutting staff at its Los Angeles divisions, and Boyd's first marriage ended in divorce. After years of working at a breakneck pace for government and industry during the hectic days of the Soviet–US Space Race, a change was in order.

Perhaps unsurprisingly, Boyd at this point pivoted back to the career that had motivated her to excel academically in the first place: that of teacher. She took up an assistant professor position at California State University, Los Angeles. This was in the middle of the revolution in mathematics teaching known as New Math (famously satirised by Tom Lehrer with the phrase, 'Where the important thing is to understand what you're doing, rather than to get the right answer'). Boyd devoted herself not only to teaching classes but also to training new mathematics teachers in the methods of the New Math, even taking up teaching two elementary classes herself as part of the

Miller Mathematics Improvement Program. In 1989, Boyd looked back on this time fondly:

> I wonder how I was able to handle a full-time load at CSULA, an evening class at the University of Southern California and the elementary school classes. I was happy in my work and I felt that I was a good teacher; hence, the full schedule was not a burden to me.

As an outgrowth of her work helping new teachers learn the New Math, she collaborated with Jason Frand to create a new textbook for teachers, *Theory and Applications of Mathematics for Teachers* (1975), which went into a second edition in 1978, and then disappeared as mathematics teaching moved away from the New Math in its perpetual quest after the Next Big Method that, surely, this time, will fix America's declining performance in mathematics.

In 1970, Boyd married real estate broker Edward V. Granville, with whom she moved to Texas in 1984, and subsequently took up a position at Texas College, which she held until her retirement in 1997. The intervening years have seen Granville honoured with medals, degrees and titles in recognition of her long service not only to the nation's space programme, but to the cause of improving American mathematical education at its most foundational, important, and often-neglected levels. As of the time of this writing, she has eighteen months to go to her hundredth birthday, and let us hope that it is a day full of even a fraction of the wonder and happiness that she has brought to students and science lovers over the decades.

FURTHER READING: In 1989, Evelyn Boyd Granville wrote a reminiscence of her life for *SAGE: A Scholarly Journal on Black Women*, which is an important source for clarifying her various changes of employment throughout the 1950s and early 1960s. To learn more about the machine that she did most of her seminal work on and the relationship between computing and NASA, you can pick up Emerson Pugh's *Building IBM: Shaping an Industry and its Technology* or James Cortada's *IBM: The Rise and Fall and Reinvention of a Global Icon* (2019).

Chapter 25

Impossible Creatures and How to Make Them: The Topological Legacy of Mathematician Mary Ellen Rudin

There's a lot to like about plain old, everyday space. No matter where you are, there's always a way to get to where you need to go, and you can always figure out how long any route from here to there might be. Calculus works in it, and that means we can solve a lot of problems about how stuff moves through it. No matter what scale you exist at, you never lack for neighbours. Yes, space is pretty great, but what, mathematically, is it about normal space that lets it be that way, and what can you take away from normal space and still have it mainly do the things we want it to?

Many branches of mathematics investigate the question of the properties of regular space, and the theoretical properties of spaces, with slightly different ground rules than the ones we are used to, but one of the most exciting of the last century and a half has been that of set topology. Here, the basic necessities for space-as-we-know-it have been minutely investigated and then, one by one, removed, to see what happens, like a multidimensional version of Kerplunk.

To be a set topologist, dreaming up new spaces with exotic properties and exhilaratingly odd behaviours, requires outside-the-box thinking of an almost unworldly degree and a genius for thinking across different mathematical landscapes as you hunt for exotic widgets and tools that combine to create a particular world with particular aspects heretofore deemed impossible. Because their work is wrapped in an almost impenetrable veil of mathematical language, the great set topologists are all but unknown to most of the world, but within the grand temple of mathematics, they are regarded with the same hushed reverence we reserve for a Shakespeare or a Marie Curie. Chief among these 'How does a brain like that even *happen*?' legends is Mary Ellen Rudin (1924–2013), a topologist who described herself as part of the 'housewife generation' of mathematics but whom other mathematicians described as an individual of breath-taking

117

insight and originality, who defined for generations to come the course of set topology.

Mary Ellen Estill was born on 7 December 1924 in the small town of Hillsboro, Texas. Her mother was a high school English teacher and her father a civil engineer whose work for the Texas Highway Department kept the family constantly on the move until the Great Depression trapped them in the rural locale of Leakey, Texas, which boasts a current population of 446 individuals. It was the sort of town that nobody ever much visited and where nothing much ever happened; where children still went to school on horseback, and spent their afternoons in imaginative play and their nights gathered around the family radio set.

Mary's childhood was a slice of Depression-era Americana, and she expected nothing more from her future than a modest teaching position and a traditional marriage. One of five graduating students in her high school class, when she went on to college at the University of Texas in Austin, her plan, encouraged by her parents, was to take a general class load that would prepare her in a non-specific way for whatever life might offer, but fate had something else in store.

In those days, to register required physically going to a central location where each department had a table set up where you could ask a faculty member what the different classes were and sign up for the ones you wanted. When Mary arrived on registration day, it was not with any particular agenda or wish list in mind, and so she was ushered to the table with the shortest queue: that of the mathematics department. Here, there just happened to be seated Robert L. Moore, one of the most eccentric and controversial figures in American mathematics. He talked with Mary about maths, about the basic elements of proof, like 'if-then' statements, and determined, based on that small interview, that Mary had a logically prone mind that needed development. Moore took her under his wing, and until her senior year every maths course she took was taught by Moore.

This was both a good and bad thing. Moore absolutely discouraged his students from reading maths papers put out by other mathematicians, and instead wanted them to cultivate their ability to solve problems from the ground up, using only starting definitions and their own innate sense for how mathematics works. The result was a generation of 'Moore mathematicians' who emerged from college with an incredible and individualistic ability to grapple with complex mathematical problems confidently, but who also possessed huge gaps in their knowledge base and no familiarity with the current mathematical literature.

At the end of her three years as an undergraduate, Estill was invited by Moore to continue at the University of Texas for graduate school, with himself as her advisor. Professionally, this was perhaps a disservice to Estill, who as a graduate student needed to be exposed to other styles and mathematical approaches, and to broaden the base of her mathematical knowledge, rather than be led deeper into Moore's idiosyncratic system with its non-conventional notations.

Though his push to have Estill stay under his supervision at the University of Texas was professionally a disservice, it was, from a human point of view, understandable. As Estill later reflected, she represented for him an opportunity that he almost never had, a brilliant and original mind as yet completely untouched by standard approaches to advanced mathematics; a mind that could be completely moulded along the lines of his principles and then unleashed to see how far it might go.

Whether he was honestly convinced she could only flourish under his guidance, or whether he was just reluctant to lose his greatest pupil, Moore ensured she stayed well within his orbit for as long as reasonably possible. When she wrote her graduate thesis in 1949, it featured the creation of a brilliant counterexample to one of Moore's own axioms, and was couched in his own virtually impenetrable mathematical symbols. It showed the great promise of her mind, but also was limited in its impact, thanks to the language in which it was couched.

When it came time for Estill to leave the University of Texas, she was told by Moore that he had already arranged for her to attend Duke the next year to teach at its women's college. Here she met her future husband, Walter Rudin, an Austrian emigrant who had fled the Nazis in the Second World War, and who was also a promising young mathematician. When they married in 1953 she took it as a matter of course that she would follow him to the University of Rochester, where he worked as a professor, and become what she called a 'housewife mathematician' – a woman who did the work of raising children and having a family, but who in her spare moments dedicated herself to mathematics and part-time teaching.

Mary Ellen Rudin had her first child in 1954, and her second in 1955, aided by an indefatigable nanny, Lila Hilgendorf, whose help gave Rudin the space she needed to pursue her intellectual goals – which included in 1955 the feat for which she is most known: the creation of a Dowker space.

Dowker spaces feature two properties that don't play well together: normality and a lack of countable paracompactness. Usually, if a space is normal (if, given two points in that set, you can always find little neighbourhoods of those points that don't intersect with each other), it is countably paracompact (given a bunch of sets that cover up the whole space, you can always create a finer cover of that space that is locally finite). Compactness and normality are two of the biggest supports for making the space we know and love work the way it does, and they tend to be fellow travellers. So, a space that is normal but not countably paracompact would be a rare creature indeed, and in 1951 Clifford Dowker theorised one could not be made, a theory which held until Rudin strolled on up and made one.

Her original Dowker space, from 1955, was based on a mathematical object which, at the time, mathematicians were not sure of the existence of: Suslin lines. This creation, which harnessed an object which might or might not exist to construct an object that had been declared impossible, did not catch on, because it existed outside the realm of the Zermelo-Fraenkel set theory (or ZFC), which forms the basis of the axiomatic system in which discussions about sets and their properties are generally couched.

Later, in 1970, Rudin would strike again with Dowker spaces, constructing a massive Dowker space within the standard confines of ZFC. Here, she made a Dowker space from a massive subspace of a particular product space under the box topology (a space's topology being defined as the way that we define 'neighbourhoods' around the points in that space, which itself affects how distance, separation and fineness are defined in that space), and thereby made the first ZFC Dowker space in two decades of trying, setting a new course for the bleeding edge of set topology.

The Rudins had moved from the University of Rochester to the University of Wisconsin in 1959, where Mary Rudin continued as a part-time teacher and researcher, but after her 1970 feat of finding a ZFC Dowker space, the university could hardly stick by the letter of old anti-nepotism laws any longer, and offered her a full professor position. Walter wanted her to take it, so that she would have a means of financial support if he were to suddenly pass away, but Mary herself was not so sure. As it stood, her work consisted precisely of those things she wanted to do – researching absurdly difficult mathematics problems and teaching now and then – without any of the things that came with being a full professor that she decidedly did not want to do – committee work, departmental politics, and the mad scramble for prestige and power. Her ideal, expressed in later interviews, was to sit in the middle of the living room in her and Walter's Frank Lloyd Wright-designed home, with children rambling about everywhere while she sat on the couch and drew her way through intriguing mathematical problems, at the centre of the family, and with plenty of time to work on maths and just maths.

She ultimately took the position, and continued to live as she had, releasing blockbuster mathematical constructions at a regular pace through the next few decades that left her colleagues dumbstruck with the originality of her mathematical inventions. Peter Nyikos, in his memorial remembrance of Rudin's work for the American Mathematical Society, expressed this sense of wonder:

> Her paper on her screenable Dowker space solved a 1955 problem of Nagami whether every normal, screenable space is paracompact. The proof of normality was a tour de force, amazing in its originality. I had never seen anything remotely like it, nor the way she was able to use the intricate set-theoretic axiom $\Diamond$++ to define the space itself. To this day I have no idea how it entered into her mind that a peculiar space like this would have all the properties required to solve Nagami's problem nor how she was able to decide on the way to use $\Diamond$++ in the definition. One part of her paper reminded me of an anecdote that was told about a session in the Laramie workshop. She had been going over a particularly intricate construction when F. Burton Jones interrupted: 'What allows you to say that?' Mary Ellen replied, 'Why that's – that's just God-given.' 'Yes,' Jones is supposed to have said, 'but what did God say when he gave it to you?'

Mary Ellen Rudin's contributions to mathematics were many, including the first non-trivial proofs about box topologies, the creation of an example that demonstrated one unproven consequence of Gödel's Axiom of Constructibility, and a proof of Nikiel's 1986 conjecture that every monotonically normal compact space is the continuous image of a linearly ordered compact space made in 1999, when she was 75 years old and had been retired for ten years. She was the reigning wizard of mid- to late twentieth-century set topology, and a respected mentor who, perhaps because of her own experience as a student having one person impose their mathematical personality on her path, was known for giving students space to work on their problems in addition to a thorough grounding in fundamental principles. She was a phenomenon, a mathematical mind unexplainable in mere terms of hard work and diligence, one of those genius presences placed on Earth from time to time seemingly to remind ourselves just what a human brain can do, and how much we have left to learn.

121

Mary Ellen Rudin died at her home on 18 March 2013.

FURTHER READING: The American Mathematical Society memoirs of Mary Ellen Rudin feature a number of mini-memorials about her work and personality that palpably demonstrate how entirely 'other' her mind appeared to her colleagues. There is also a more through-composed biography and appreciation of Rudin to be found in Claudia Henrion's *Women in Mathematics: The Addition of Difference* (1997), while Rudin's most famous work, her ZFC Dowker construction, can be found online.

Chapter 26

Non-Linear: How Mathematical Lone Wolf Karen Uhlenbeck Found Her Pack

When you first walk into secondary school your first year and plop yourself nervously into a desk in the back of your geometry class (because you're *far* too cool to sit in the front seats), one of the earliest things you're asked to do is to grapple with geometric proofs. For many, it is a gruelling experience soon forgotten, but for a few it is a beautiful initiation into a world of rigour and logic, wherein you proceed from point A to point B with the inexorable accumulated force of two millennia of mathematical insight. It is a seductive vision, so neat and ordered and linear, and that too-clean picture of the guts of mathematics has driven more than a few promising students out of the field, racked with self-doubt: *Maths is an ordered march from unknowns to new theorems that should be crystal clear in its origins and elegant in its ultimate formulation. I don't think that way, so I guess I shouldn't be in maths.*

Certainly, there are many examples of great mathematicians who were every bit the lucid, logical, linear thinkers of the classic mould, but we are doing maths and science a disservice whenever we fail to point out that there are just as many who aren't. There are many roads to mathematical success, as long as you possess the courage to maintain your identity in the face of struggle, and the wisdom to seek support in a larger community during your growth.

Now that seems common enough advice, but it was decidedly not in the 1950s when a young woman named Karen Keskulla (b. 1942) set out to discover who she was, and what she was good at. The daughter of an engineer and an artist, she grew up a tomboy and book nerd, spending her free time playing football with the boy up the street and her class time secretly reading advanced science books under her desk.

She loved roaming through the local countryside and shutting herself up with the written word, and dreamed of a job that would leave her to herself. 'I was either going to be a forest ranger or do some sort of research in science. That's what interested me. I did not want to teach. I regarded anything to do with people as being sort of a horrible profession.'

To that stereotype that says (and you can sing along if you have heard the tune before), 'Women don't do maths because, biologically, they are more social than men and therefore unhappy in careers that require isolation and lone research,' Keskulla is a great counterexample proving how dangerous it is to generalise about what women are Biologically Like. She defied gender roles, even as a young girl at the start of the Eisenhower Era who knew nothing so much as that she wanted space and time to herself to work alone on the big problems of nature.

'I never had very many boyfriends; I didn't feel comfortable. I never felt like I was really a part of anything. I went my own way, without really wanting to, but I never did understand the trick of doing things like you were supposed to.'

She went to the University of Michigan to study physics but switched to mathematics when she found it suited her style and interests, but her decision to follow up her promising beginning with graduate work was made almost as much for social as intellectual reasons. Most of the people in her social circle were heading to graduate school, including her boyfriend, and even though she had reservations about Fitting In at a graduate level maths department, she went to the Courant Institute and then Brandeis University to earn her master's and PhD. She had purposely avoided applying to more prestigious schools like Princeton or Harvard, feeling that the pressure exerted by an overwhelmingly male academic culture would distract her from her work by putting her constantly on edge.

She roved from topic to topic, watching again and again as her initial excitement for a new realm of mathematics tapered off, wondering if she would ever find a research area that could engage her fully for an extended period of time. Meanwhile, she married and in becoming Karen Uhlenbeck had to face a whole new set of professional hurdles. Her husband, a biological physicist, had offers from Stanford and Princeton, but those universities refused to hire Karen, citing nepotism laws publicly while privately acknowledging a simple reluctance to hire a woman for a mathematics position.

Sportingly, Mr Uhlenbeck declined to work at any institution that wouldn't also find a position for his wife, and the two ended up at the University of Illinois at Urbana-Champaign for five of the worst years of Uhlenbeck's life.

At Urbana-Champaign she was treated as a faculty wife rather than as a professor in her own right, and was expected to act socially in accordance with that position. She felt isolated and undervalued, lacking both a professional support network and a personal guide on how to maintain her self-worth in the teeth of her unique position. She didn't like teaching, struggled to find a topic of research that engaged her, and was socially battered by the non-recognition of her status and value. In 1976, she left for the University of Illinois at Chicago.

She had also split up with her husband, meaning that she was entirely on her own in a new environment, either to repeat the cycle of the past or to find her way at last. Fortunately, Chicago was a thoroughly different environment from Urbana-Champaign. There were other women professors there to support and advise her in her career, and other mathematicians who took her seriously and whom she could use as a sounding board in the development of her own work.

She also received a Sloan Fellowship, which allowed her to financially support herself. Now, that sounds like an 'Oh that's nice' kind of fact but it is actually one of those *critical* things we don't talk about enough in the history of women in science – how access to fellowships, grants and equal pay affects the willingness of women to continue in a career that might be socially hostile. As it was, the Sloan Fellowship gave Uhlenbeck reassurance that she could continue in her research while still supporting herself, and that little bit of extra stability turned out to be a key component of her resolve to move forward in a career that had been thus far on the frustrating side.

At Chicago, Uhlenbeck, the loner kid who dreamt of a job as far removed from other people as possible, learned that her tendency to isolate herself only hurt herself in the long run, both professionally and emotionally. She found the value of an academic support system, and encouraged her students to not fall into the old mathematician trap of holing up in a room and working yourself mad, but rather to form connections and relationships with co-workers, to be a different sort of mathematician better able to weather the rough times that inevitably come.

Settled and supported at last, she could turn her full attention to the research that would place her in the top rank of American mathematicians. In 1982, she published two papers, 'Removable Singularities in Yang-Mills Fields' and 'Connections with L^p Bounds on Curvature' that paved the way for groundbreaking work in gauge theory and the nature of exotic n-spaces.

That is a lot to elucidate, but essentially her work is part of a mathematical continuum dating back to 1919 when Herman Weyl proposed taking Einstein a step further. Einstein had shown, in his theory of general relativity, how to compare measurements taken by two observers at different locations in a gravitational field. In special relativity, measurements taken by different observers can be simply related to each other through a Lorentz transformation, but when you throw in significant differences in the gravitational field, the comparing of measurements becomes trickier. Einstein's general relativity solves this problem by employing 'connections' between observers based in the geometry of space-time.

Weyl wondered if he could do the same for electromagnetism. Thus began the field of gauge theory, in which we look for the connections that bind and compare measurements taken at different points in a field. This discipline of applied mathematics was bedevilled over the coming decades by a plague of Too Earlyism. Weyl's insights came too soon, and had to wait until the era of quantum physics to see their first lasting results. In 1954, Yang and Mills attempted to push the model further to shine light on the nature of the strong force, only again to run against difficulties that would take a decade and more to clear up.

The end result of this half century of mathematical bundles, sweat and fibres was the unification of the electromagnetic force with the weak force (which governs processes of nuclear decay), through the analysis pioneered by Weyl and cemented by Yang and Mills. Uhlenbeck's most noted work, then, focused on the application of gauge theory to four-dimensional manifolds (basically, objects that look up close like normal Euclidean four-dimensional space but that might not look that way from afar). She and C.H. Taubes analysed Yang-Mills equations in four dimensions, laying the groundwork for the theories of Simon Donaldson, who would win the prestigious Fields Medal in 1986 for his extension of their work.

Uhlenbeck has not won a Fields Medal, though she was given the National Medal of Science in 2000, the American Mathematical Society's Leroy P. Steel Prize in 2007, and made history in 2019 when she became the first woman to win the Abel Prize for her 'achievements in geometric partial differential equations, gauge theory and integrable systems'. She is also on anybody's shortlist of the most important American mathematical researchers of the twentieth century, a beloved advocate of greater gender diversity in STEM, and a dedicated advisor attempting to create a generation of maths students less prone to implosion. From a girl who sought isolation, she has become a load-bearing pillar of both modern mathematics and the broader scientific culture it nests within.

FURTHER READING: Uhlenbeck has a chapter devoted to her in Claudia Henrion's indispensable *Women in Mathematics: The Addition of Difference*, and since winning the Abel Prize the number of online resources devoted to her life has significantly expanded. If you're interested in going deeper into the history of gauge theory, you have got a long road ahead but if you have already had your standard calculus courses, a college-level linear algebra course, some classical analysis, and some familiarity with differential equations you can head to Andrew Pressley's excellent *Elementary Differential Geometry* (2001) to start wrestling with problems of curved surfaces and how to compare them locally to Euclidean space, and once you're pretty comfortable in that, K. Moriyasu's *An Elementary Primer for Gauge Theory* (1983, reprinted 2009) is a good accounting of the history of gauge theory along with a mathematical exploration of some of its most important results in electroweak theory, the strong force, and beyond!

Chapter 27

Expectations Defied: The Algebraic Journey of Raman Parimala

It is an all-too-recurrent narrative sequence in the history of Women in STEM:

(1) Brilliant woman researcher establishes herself, with a good reputation and a solid position at an institution of renown.
(2) Brilliant woman researcher gets married to a man whose work is in a different city/country.
(3) Brilliant woman researcher moves with husband to that new city/country, where there are no significant jobs in her field, and she spends the subsequent decades doing part-time work on the fringes of her discipline, which proceeds to pass her by without a second thought.

So often do we see this pattern in the history of women in science, that we have come to expect it, and gird ourselves instinctively for the coming tragedy whenever we come to the part of a scientist's biography that mentions their marriage to Eminent Professor X or Travelling Salesman Y. But every so often, the best instincts of human nature win, genius is recognised, and our expectations are subverted by spouses who place the support and nurturing of their wives' profound gifts ahead of their own egos or immediate career goals.

This was decidedly the case for the husband of the brilliant algebraist Raman Parimala (b. 1948), who, when he saw how relocating had hurt Parimala's mathematical career, undertook to give up his job and move to the city that would best support her continued world-class work. As remarkable as that sacrifice was, however, it was not Parimala's first experience with the goodness of humanity standing out against the inertia of societal expectations.

She was born on 21 November 1948 in Mayiladuthurai, a town approximately 175 miles distant from Chennai in the Indian state of Tamil

Nadu. This was a fateful time in India, which had achieved independence from Great Britain just sixteen months before Parimala's birth, and which bore the recent scars of the turmoil that resulted from the India–Pakistan partition of 1947. Passions were high, and while many doubled down on religious and ethnic identities, and the traditional roles associated with them, others saw the transition as an opportunity to create a more equitable future, free from the restrictions of the past. Parimala's father, a professor of English Literature, was one such figure, and he actively encouraged his daughter's passion for mathematics, sending her to the Sarada Vidyalaya Girls' High School in Chennai. This was one of the institutions founded by the great Indian social reformer R.S. Subbalakshmi (1886–1969), who spent the early years of the twentieth century establishing organisations that fought against some of the reigning restrictions on women's education and social equality, including the Lady Willingdon Training College and Practice School, various ladies' clubs, schools for adult women, social welfare centres, and the Sarada Vidyalaya (founded either 1921 or 1927).

Subbalakshmi was married while she was still a child, but the early death of her husband left her free to pursue her education at the university, and she fought the rest of her life to make sure other young women in India had the same opportunity. Parimala benefited from the pioneering efforts of Subbalakshmi, received an excellent education, and encouragement to continue on with her studies, attending Stella Maris, a women's college which had only been founded in 1947, where she received her bachelor's degree in 1968, and her master's in 1970. While at Stella Maris, her goal was to simply get her degree and move on to a teaching position at the college, but one of her teachers there, Professor Thangamani, believed so heavily in her mathematical brilliance that she refused to let the college consider Parimala's teaching application, effectively shoving her for her own good into applying to doctorate programmes elsewhere.

Parimala began her doctoral career at the Ramanujan Institute of the University of Madras, but after one year transferred to the Tata Institute of Fundamental Research, a scientific research institution founded in 1945 working under the umbrella of the Department of Atomic Energy. She worked in the group of the algebraist Ramaiyengar Sridharan (b. 1935) and was soon displaying mathematical abilities in the realm of algebra that startled her older, more experienced co-workers. She was all set up to join the Tata faculty upon receiving her PhD in 1976, when she married a man who worked as the chief internal auditor for the Board of Internal Trade in Tanzania, a country which itself had just emerged into independence a decade and a half earlier. While Tanzania was more stable than other former colonial territories in the 1970s, under the comparatively idealistic dictatorship of Julius Nyerere, its capital of Dar-es-Salaam, where Parimala moved with her husband in 1976, was not precisely a world capital of mathematical research.

Her career suffered in the absence of the colleagues and robust research libraries that were the great necessities of a mathematical researcher in the pre-Internet age, and her husband was not long in noticing that Parimala's mind was going to waste in this new environment. She was offered a prestigious post-doctoral position at the University of Lausanne in Zurich, Switzerland, an institution that had been in continuous existence since 1537, and her husband agreed to give up his post in Tanzania so that she could take the job. This was the start of her meteoric rise to the first rank of the world's algebraists, anchored by her early feat of developing the first example of a non-trivial quadratic space over an affine plane. (A quadratic space is basically a vector space V, and a set of elements A, with a rule Q for mapping elements of V to a single element of A which states that, for any v in V, and a in A, $Q(av) = a^2 Q(v)$. Parimala found an example of one where A is an affine plane, i.e. a plane that is like a Euclidean plane, but has been stripped of all notions of measuring length or angle size.)

Since that impressive entry into the consciousness of the global mathematical community, Parimala has studied a variety of topics in algebra, toppling long-standing conjectures and solving others. She has worked with Witt groups, the Hasse principle, Galois algebras, torsors, Clifford algebras, and most recently with the Grothendieck-Serre Conjecture. After her time at Switzerland, she returned to the Tata Institute, where she carried out her research until moving to Atlanta's Emory University in 2005, a move that gave her more opportunities to teach the next generation of mathematicians, and which placed her closer to her son in New York. In 2010, she was made the plenary speaker for the International Congress of Mathematicians, a signal honour capping four decades of brilliant work across the algebraic domain.

Today, aged 73, she continues to teach at Emory while serving as a judge for some of mathematics' most prestigious prizes, including the Infosys Prize, and the Abel Prize. In 2020, she was recognised by the government of India as one of eleven Indian women scientists who would have a Chair named in their honour, cementing forever in the memory of her country the name of the woman who, born during the heady, chaotic days of her nation's modern becoming, went on to embody in her own person the spirit of that new country, full on the promise of its coming future, enhanced by the participation of all its people.

Chapter 28

The Billion Roads from Here to There: The Graph Theory Combinatorics of Fan Chung

'Well, some go this way, some go that way. But as for me, myself, personally, I prefer the short-cut.' – The Cheshire Cat

Every morning we wake to a deafening barrage of choice, decisions that crackle and branch to more choices and more decisions that daunt enumeration. How can we tally the untaken roads of a day, or quantify the chains of action that we might just as easily have been nudged into treading? What is the mathematics that controls the probabilities of where an initial system, governed by a complicated set of allowable connections, ends up after a day or a century?

These are the concerns of graph theory, an enchanting branch of mathematics charged with investigating how massively interlinked systems of points behave, and what to expect when you head out on a random journey from one of those points. And few people since Paul Erdös have wielded that mathematics, and enhanced its connections with other parts of existence from manifold theory to the connectivity of the entire Internet, as Fan Chung (*b.* 1949).

She is such a Maths Person, so wrapped up at every moment of the day in solving new problems and fine-tuning old solutions, that I would not remotely be surprised to learn that she was immaculately conceived by mathematically inclined midi-chlorians. That, I am sad to say, is not the case. She was, in fact, born in Taiwan to a family who encouraged her, against tradition, to pursue a life of science and in particular, given a certain lack of manual finesse when young, a life of calculation where she had to handle nothing more fragile (or explosive) than a pen and a sheet of paper.

Chung went to National Taiwan University and was drawn to the allure of combinatorics, the branch of mathematics that, broadly speaking, asks questions about the number of ways that a given event can happen. How many ways are there to arrange eight books on a shelf? How many different licence plates can you make if the first three positions are digits, the last three are letters, and no repetition is allowed? What is the maximum number of possible distances between n points in a plane?

That attraction was fostered by Herbert Wilf at the University of Pennsylvania, where Chung studied after graduating from NTU in 1970. Her score on the university's qualifying exams was so much higher than that of any other student he decided to snatch her up and make her into one of his combinatoric warriors. He lent her a book that contained a chapter on Ramsey theory so charming and enticing that it was his standard text for hooking promising students into the field. She came back in a week not only having read and understood the work, but having found a spot that could be improved. 'In just one week,' he recalled, 'from a cold start, she had a major result in Ramsey theory. I told her that she had just done two-thirds of a doctoral dissertation.'

It was the first step in a string of triumphs that would encompass hundreds of papers producing important results in half a dozen different fields of mathematics, none of which will make the least bit of sense without at least a few words about what graph theory is all about.

First of all, it is not about graphs. Well, not graphs as we know them from school, with their curvy parabolas and pointy absolute value functions jauntily marching about in the space created by the x and y axes. A graph $G(v,e)$ is simply made up of v vertices connected by e edges. Here's an exquisite, some might say artisanal, drawing of a random $G(4,5)$ graph, for example:

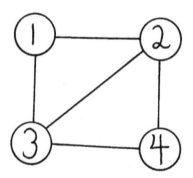

In practical life, you can think of the points as oil reserves connected by pipelines, or websites connected by links, or people connected through a social network, at which point seemingly abstract questions in graph theory such as, 'What is the minimum number of points we need to remove to cut the graph into two disconnected pieces?' or, 'If we let a set of free agents wander this graph, is there a stable final configuration, and how fast do we get to it?' become multi-billion dollar questions pushing the future of global commerce and information technology.

One of the areas that highlights all of the different areas of mathematics that Chung brings to her research on combinatorics and graph theory involves questions of random walks along a graph. Put simply, if you put a number of people at the different vertices of a graph and let them walk from vertex to vertex however they want, is there a stable configuration that they'll eventually end up at, and how long will it take them to reach it? Stated another way, if each of the vertices is a web page, and people randomly click on links, is all the clicking going to settle into some stable number of people we expect to be, at any given moment, on any given site?

Let's do it, then! Suppose we have our G(4,5) graph above and we put one person at each corner, then let each of them randomly choose an edge to walk along towards another corner. So, the person at corner 1 can walk to corner 2 or 3, but the person at corner 3 can walk to corners 1, 2, or 4. The initial configuration can be expressed as follows (i.e. with just one person at each vertex):

$$(1 \quad 1 \quad 1 \quad 1)$$

Now, we need to make a matrix that expresses the probability that a person will choose to go from corner i to corner j. It is called a transformation matrix, and here it is!

To Vertex:	1	2	3	4

$$From\ Vertex \quad \begin{matrix} 1 \\ 2 \\ 3 \\ 4 \end{matrix} \begin{pmatrix} 0 & 1/2 & 1/2 & 0 \\ 1/3 & 0 & 1/3 & 1/3 \\ 1/3 & 1/3 & 0 & 1/3 \\ 0 & 1/2 & 1/2 & 0 \end{pmatrix}$$

We have 1/2 as the entry in the 2nd column of the 1st row because a person at vertex 1 has a 1/2 chance of going to vertex 2. But the 1st column entry of row 3 is a 1/3 because, from vertex 3, I have three choices of where to go so the chance of going to vertex 1 is just 1/3. So, to figure out how everything ends up after everybody's first random choice, we just have to multiply the original configuration matrix by the probability matrix to get the new configuration:

$$(1 \quad 1 \quad 1 \quad 1)\begin{pmatrix} 0 & 1/2 & 1/2 & 0 \\ 1/3 & 0 & 1/3 & 1/3 \\ 1/3 & 1/3 & 0 & 1/3 \\ 0 & 1/2 & 1/2 & 0 \end{pmatrix}=(2/3 \quad 4/3 \quad 4/3 \quad 2/3)$$

After one decision we see that people are starting to group up around vertices 2 and 3, which is what we'd expect, since there are three roads to each of those vertices but only two roads to vertices 1 and 4. Now, to see what happens with the next random move, we just need to multiply our new distribution of people by our old friend the transformation matrix, which thankfully never changes!

$$(2/3 \quad 4/3 \quad 4/3 \quad 2/3) \begin{pmatrix} 0 & 1/2 & 1/2 & 0 \\ 1/3 & 0 & 1/3 & 1/3 \\ 1/3 & 1/3 & 0 & 1/3 \\ 0 & 1/2 & 1/2 & 0 \end{pmatrix} = (8/9 \quad 10/9 \quad 10/9 \quad 8/9)$$

The question, then, of 'Does it settle down?' can be answered by just continuously multiplying by the transformation matrix until the numbers stop significantly changing. Try it at home and you should get (.8 1.2 1.2 .8) as your stable configuration. If we had 100 people at each corner, then, we'd expect the stable state to have 80 each at corners 1 and 4, and 120 each at 2 and 3. That's fine, but it doesn't answer questions like, 'How quickly does the system achieve this stable state?' It makes rather a big difference if it will settle to these proportions after thirty choices or after 30 million, especially if this graph models something physical.

Enter eigenvalues. This is where the 'spectral' in Spectral Graph Theory comes from. It is a rather alarming word for a really simple idea. Sometimes, when you multiply a matrix by a one-column matrix, the result is just a multiple of that one-column matrix. The multiple is called the eigenvalue and the one-column matrix is called the eigenvector. A matrix's eigenvalues carry with them massive amounts of information about the matrix they come from and, therefore, when that matrix represents a physical model, about the physical situation it represents.

We're interested in the eigenvalues of a matrix called the Laplacian, which is relatively easily built:

- Put 1's along the main diagonal.
- For the entry in row i, column j, put a zero if vertex i and j aren't connected by an edge, and put $-1/\sqrt{(d_i d_j)}$ if they are, where d_i and d_j just represent the number of edges coming out of vertices i and j. For those playing along at home, our graph from above would have the following Laplacian matrix:

The Laplacian, yo!!

$$\begin{pmatrix} 1 & -1/\sqrt{6} & -1/\sqrt{6} & 0 \\ -1/\sqrt{6} & 1 & -1/3 & -1/\sqrt{6} \\ -1/\sqrt{6} & -1/3 & 1 & -1/\sqrt{6} \\ 0 & -1/\sqrt{6} & -1/\sqrt{6} & 1 \end{pmatrix}$$

Find out the eigenvalues of THAT matrix, and it turns out you find out a lot about how quickly the graph converges to its stable state. Now, that all looks simple – an incredibly clever use of methods you learned in your pre-calculus class or undergraduate linear algebra course – but that is because we're dealing with one given case with a very small number of vertices. It is when you start saying general things about systems of n vertices and e edges that life gets interesting, and it is precisely those questions that Fan Chung has been shedding light on for the better part of five decades now, creating solutions for general values of n that have allowed us to gain insight into the comings and goings of massively complicated systems like website connectivity on the Internet or the chains linking members of a biological network.

How do you deal with random paths when the different edges have different weights? How does giving each edge a direction (i.e. you can only go from site A to site B, never site B to site A) mess with the ability of a system to settle down? First at the University of Pennsylvania then, throughout the 1970s and 1980s, as a researcher and manager at Bell Laboratories and its splinter company, Bell Communications Research, Fan Chung has tackled the intersection of abstract graph theory and complex real-life applications by collaborating with mathematicians in far-flung fields, to bring unlikely approaches to solve sticky problems. She describes mathematicians as falling into the category of either Theory Makers or Problem Solvers, and places herself firmly in the second category, as someone who thinks about maths every waking moment, getting better answers to old questions and finding ways to apply several disciplines' worth of wisdom to answer the questions posed by the exploding world of Big Connectivity.

And the maths and the life keep rolling on. She has had two children, working until the very last moment while pregnant with each and even continuing her research at home when she took a few weeks' personal vacation after the births. Her husband is a mathematician in the same field, and their joint passion for constant research draws them together in investigation, rather than pushing them apart in competition. Where other couples have a massive television sprawled along a wall of the living room, they have a giant whiteboard where they solve problems together, their work, their passion, their hobby all concentrated on the same intellectual object.

It is, by all objective standards, a magnificent way to spend a life with another human.

Today, Fan Chung's research continues at UCSD, where she has been a professor since 1998, and as the editor-in-chief of *Internet Mathematics* since 2003. She manages the 'bigness' of the virtual world we have placed ourselves in, using the massive power of graph theory to find how tribal online communities work in isolation, how they interact with the larger picture, and what the effects might be should they be suddenly expunged, while documenting the growing disparity between high-volume, high-connectivity sites and the small competitors treading water ever more feebly in their wake.

Maths might be pretty, but it is rarely kind.

FURTHER READING: Chung has two books out, *Spectral Graph Theory* and *Erdös on Graphs*, both of which are pretty easy to find, but I think are probably too daunting for most. Better to start with Béla Bollobás's *Extremal Graph Theory* and work your way up, supplementing it with the fantastic resources that Chung has on her website, including her lecture notes, which clarified beautifully a number of points that I had been struggling with when trying to grasp the intersection of spectra with graphs.

Chapter 29

It Came from Teichmüller Space! The All-Too-Brief Mathematical Adventures of Maryam Mirzakhani

A square, who works as a lawyer in the two-dimensional world of Flatland, sits down with his hexagonal grandson:

> Taking nine squares, each an inch every way, I had put them together so as to make one large square, with a side of three inches, and I had hence proved to my grandson that – though it was impossible to see the inside of the square – yet we might ascertain the number of square inches in a square by simply squaring the number of inches in the side: 'and thus,' said I, 'we know that 3_2, or 9, represents the number of square inches in a square whose side is 3 inches long.'
>
> The little hexagon meditated on this a while and then said to me: 'But you have been teaching me to raise numbers to the third power: I suppose 3_3 must mean something in geometry. What does it mean?' 'Nothing at all,' replied I, 'Not at least in Geometry; for Geometry has only Two Dimensions.' ... My grandson, again returning to his former suggestion, exclaimed, 'Well, if a Point by moving three inches, makes a Line of three inches represented by 3, and if a straight Line of three inches, moving parallel to itself, makes a Square of three inches every way, represented by 3_2; it must be that a Square of three inches every way, moving somehow parallel to itself (but I don't see how) must make Something else (but I don't see what) of three inches every way – and this must be represented by 3_3.'
>
> 'Go to bed,' said I.

This excerpt, from Edwin Abbott's lusciously nerdy 1884 satire *Flatland*, was written on the eve of Einstein's space-time revolution, and captures

137

nicely the common-sense anxiety of casting one's imagination beyond the space you happen to live in. Over a century later, four dimensions are the least of our mathematical worries, and the way forward is lit by our own irrepressible human hexagons – people with the knack for peering into abstract spaces and wresting from them consistent laws. And of all our daring hexagons, few ranked higher than the first woman to win the Fields Medal, Maryam Mirzakhani (1977–2017).

Mirzakhani was a scribbler of the first order – a kinetico-visual thinker who filled vast sheets of paper with sketches probing at the edges of maths' biggest problems. By the age of 36, she had already solved enough of pure maths' Insoluble Enigmas to fill two careers, and her pace seemed to show no sign of slowing down to slouch over past greatness, when suddenly a diagnosis of breast cancer in 2013 sharply hemmed in that boundless horizon of mature potential behind mute walls of steady menace.

Born in Tehran in 1977, just two years before the Revolution that set that country on a new and challenging course, Mirzakhani was, from the first, a courter of the unlikely. A daydreamer and bookworm, writing seemed a natural choice. Her broad and lively capacities were recognised early, and she was diverted into a school run by the National Organization for the Development of Exceptional Talent where, after a slow start, she soon became the star mathematical pupil, winning the Mathematical Olympiad gold medal two consecutive years, in 1994 and 1995.

That led to an undergraduate degree at Sharif University, and thence graduate work at Harvard, where she produced her first mathematical masterpieces. These papers dealt with hyperbolic surfaces and moduli spaces. And that is where we get into some clever stuff.

The story of hyperbolic surfaces is, really, one of the oldest tales that mathematics has to tell. It all begins with the Axiom That Wasn't, Euclid's Fifth. From its birth, mathematicians found it an odd duck, a statement that didn't quite seem to fit with Euclid's other foundational assertions. Expressed in modern terms, it simply says that, if you give me a line and a point not on the line, then there exists exactly one unique line through that point which is parallel to the original line.

And so there is – as long as the space where those objects live happens to be flat. So evident does it seem, that mathematicians spent entire careers trying to bend geometry to make it fit naturally in the position that Euclid gave it. To no avail. Finally, after centuries of time-wasting, it was realised that one could, in fact, construct geometries where The Fifth was not true, one of which was the hyperbolic plane, most easily visualised, I think, through the version known as the Poincare half-plane, sketched below:

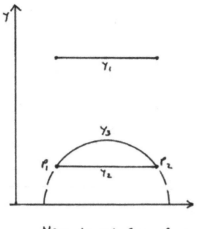

Hyperbolic Distance = $\dfrac{\text{Euclidean Distance}}{Y}$

So, closer to The bottom, hyperbolic distances get bigger. Y_1 & Y_2 have the same _Euclidean_ distance, but Y_2 has a much greater _hyperbolic_ distance.

· Now, to get from P_1 to P_2, we _don't_ want to take Y_2 — that goes through some very _thick_ space. The shortest distance path is actually the circular arc, Y_3.

In this world, all space is not created equal. It gets, in essence, thicker as you move closer to the bottom of the half-plane. My hyperbolic geometry teacher used to tell us to think of it as having the consistency of honey near the bottom axis – hard to move through – and getting progressively easier to navigate as you moved upwards. As such, the quickest way to get from one point to another directly to the right of it is NOT the straight segment that connects them (y_2 in the figure) – that way you'd be running through the thickest space the whole trip. Far better to head upward, where the path is a bit easier, and then to loop your way back downward. And, in fact, the path of shortest distance between these two points, called a geodesic (remember that word), lies on the semicircle through them which has its endpoints on the bottom line (y_3 in the figure).

So, since lines in this world are semicircles, it is possible, if you give me a semicircle and a point not on it, to construct more than one semicircle through that point that does not intersect the original semicircle, and therefore is considered parallel to it. This space obeys Euclid's first four postulates, but breaks the Fifth, and introduces a slew of new geometric possibilities.

A hyperbolic *surface*, then, is a metric space (a space with a way to measure distance) where, if you take a neighbourhood around any point, it is related to a neighbourhood of points in the hyperbolic plane we just talked about. Such a surface contains all the craziness of the original hyperbolic plane, kicked up to the next level. These surfaces, understandably, inherit some rather interesting geometry, and it was Mirzakhani's task to tame the chaos. In particular, she wanted to break the mystery of how many simple closed geodesics of a given length a hyperbolic surface possesses.

Put more plainly, how many shortest paths of a given length are there that form a closed loop without intersecting? Let's stop and appreciate how intense that question is. It is asking for a method to determine, for an object that cannot exist in real space, with geometry inherited from a brilliant non-Euclidean dodge, with geodesics ranging from the infinite to the well-behaved, how many of a given length there are going to be, which don't cross themselves, which end where they begin.

Insane. But Mirzakhani did it, and that was just part of what she accomplished as a *grad student*. From there, she studied, first at Princeton (2004–2008) then Stanford (2008–2017) the world of moduli spaces, which are harder still to grasp. Oversimplifying egregiously, a moduli space is a space where each point represents some mathematical object or class of objects. Mirzakhani's research focused on Teichmüller spaces, which are closely related to the Riemann moduli space. Basically, to get a Teichmüller space, take a surface, let's call it X, and make sets called 'complex structures' of equivalent maps between that surface and the Euclidean plane. Doing just that lets you construct the Riemann moduli space of X, but if you add one more requirement about what it takes to call two structures equivalent, you get Teichmüller space.

In other words, a Teichmüller space is a space where each point represents a whole class of equivalent complex structures. That is a rather abstract mental world to live in, but then to think about what happens when you put a strain on that system is something else entirely. Mirzakhani's work considered what happens as geodesics are made to flow along a Teichmüller space, discovering that the phenomenon has ergodic properties. That realisation brought a whole new realm of tools to bear on the problem, and broke it elegantly from Impossible Conundrum to Solved Case.

And it didn't end there. Work on billiard reflection with Alex Eskin resulted in a paper that opened up brand-new sprawling fields of mathematical research. As ever in mathematics, work in the physical world led to new abstract results, which themselves lead to entirely unexpected physical ramifications. Mirzakhani's career seemed set for superstar status, particularly after she became, in 2014, the first woman to win the Fields Medal, but within two years of that historic win, her doctors discovered that her cancer had spread to her liver and her bones, and that her time on this planet had grown perilously short. She was a brave and brilliant hexagon who gave the full sum of her genius to allowing the rest of us squares to comprehend, if just tentatively, the hidden structure of the abstract world she saw as clearly as we do the birds and the sky, and when she passed away on 14 July 2017, in California, the mathematical world lost a little of its sparkle, and its future more than a little of its promise.

FURTHER READING: If you want to start getting into this area of mathematics, and have had the usual upbringing in maths, a good place to start is *Topology* by James R. Munkres. It gives you the framework for thinking about open sets, mappings, and all the good stuff you need to think about what happens as we cut and paste the edges of reality in new ways. For the hyperbolic plane, I like Saul Stahl's *The Poincare Half-Plane: A Gateway for Modern Geometry*. It develops the Euclidean stuff at a good pace before having you jump into the hyperbolic material and requires really just basic calculus and trigonometry.

Chapter 30

Brief Portraits: The Modern Era

Mary Emily Sinclair (1878–1955)

Sinclair's mathematical work centred around the subjects of algebraic surfaces (objects whose points form a two-dimensional fabric that contains all the solutions to a given polynomial equation), and calculus of variations (the branch of mathematics concerned with finding what function maximises or minimises a certain quantity under certain restrictions – for example, of all the curves that connect two points A and B, which one would take the least time for a particle to travel along under the influence of just gravity?). She was born on 27 September 1878, in Worcester, Massachusetts, where her father worked as a mathematics professor at the Polytechnic Institute.

She was the middle child of three daughters, all of whom were sent to college. Sinclair attended Oberlin College, graduating in 1900, and did her graduate work at the University of Chicago, earning a master's in 1903 for her thesis on the classification of algebraic surfaces formed by the solutions to the equation $u^5 + 10xu^3 + 5yu + z = 0$, and a PhD in 1908. Her dissertation, 'Concerning a Compound Discontinuous Solution in the Problem of the Surface of Revolution of Minimum Area' focused on a problem in the calculus of variations whereby you are trying to determine – given two points on the same side of the x-axis connected by an arbitrary curve, and a third point on that arbitrary curve connected to the x-axis by another arbitrary curve – what arbitrary curves will produce the surface of least surface area when rotated around the x-axis.

She began teaching at Oberlin in 1907, and was promoted to associate professor in 1908 following the completion of her dissertation, and then to full professor in 1925. In 1914, she adopted a baby girl, Margaret, and in 1915, a son, Richard, taking time off from Oberlin in 1914, 1922, and 1925 to study at Columbia, Johns Hopkins, the University of Chicago, Cornell, the University of Rome, and the Sorbonne. With the coming of the Great Depression, which brought with it a substantial cut to her salary, and lingering problems with her daughter's health, she left off taking lengthy sabbatical studies until 1942, and retired in 1944. In 1950, she was heavily beaten during an attempted carjacking, and moved into a house with her daughter-in-law in Maine, where she died in 1952.

Lennie Phoebe Copeland (1881–1951)

Born in Bangor, Maine, on 30 March 1881, Lennie Phoebe Copeland received her bachelor's from the Educated University of Maine in 1904, then taught at high school for five years before attending Wellesley in 1911 to receive her master's, and the University of Pennsylvania in 1913 to receive her doctorate for a dissertation on the invariants of n-lines (invariants are characteristics of a mathematical object that remain unchanged when that object is put through a particular transformation, while n-lines are essentially plane polygons). She then returned to Wellesley in 1913 to teach, reaching the rank of professor by 1937, and professor emerita by 1946, continuing in that role until her death in 1951.

Anna Pell Wheeler (née Johnson) (1883–1966)

The daughter of Swedish immigrant parents, Anna Johnson was born on 5 May 1883, in the charmingly named town of Calliope, Iowa. When she was 9 years old, her family moved to Akron, Iowa, where she attended high school. She attended the University of South Dakota in 1899, receiving her bachelor's degree in 1903. One of her teachers at USD was Alexander Pell, a former Russian double agent and revolutionary who had been compelled to flee Russia. He encouraged Johnson to continue her studies at the graduate level, whereupon she received master's degrees from the University of Iowa in 1904, and Radcliffe in 1905, before taking the now-obligatory pilgrimage to the University of Göttingen in 1906.

Pell joined her there, and in 1907 the teacher and his former student were married in Göttingen. Johnson completed her thesis, 'Biorthogonal Systems of Functions', but did not receive a degree due to some unknown disagreement with Hilbert. The paper was acceptable, however, to the University of Chicago, which awarded her a PhD in 1910, and was published in 1911. (Biorthogonal systems are made up of two spaces of indexed vectors, equipped with a mapping that takes a vector from each space and outputs 1 if the two vectors are equal, and 0 if they aren't.)

Pell suffered a stroke in 1911, and for the next ten years, until his death in 1921, Johnson had to balance teaching at first Mount Holyoke College and then Bryn Mawr, her own mathematical research (including a 1917 paper on Sturm's Theorem), and overseeing Pell's medical care. In 1925, Johnson remarried, this time to a Bryn Mawr Classics professor, Arthur Leslie Wheeler, who died in turn in 1932. Johnson continued teaching at Bryn Mawr until 1948, from which position she directed the effort to assemble an asylum package for Emmy Noether that allowed her to flee Nazi Germany in 1933.

Pauline Sperry (1885–1967)

Sperry is known today mainly for her refusal in 1950 to sign a McCarthyist loyalty oath and subsequent dismissal from UC Berkeley, and her reinstatement, with back pay, in 1952, when the US Supreme Court ruled that dismissal unconstitutional in Tolman v. Underhill. But this is to ignore her role as an early figure in American projective differential geometry, in which field she made several contributions in the 1910s and 1920s. She was born in Peabody, Massachusetts, to two teachers, her father eventually becoming president of Olivet University, which Sperry attended before switching to Smith College, where she studied mathematics and music, receiving her bachelor's degree in 1906. Entering the University of Chicago in 1913, she came under the mathematical influence of Ernest Wilczynski, who was the founding figure in American projective differential geometry, which studies properties of curves that remain unchanged under projection. Pauline Sperry's master's thesis, 'On the Theory of a One-to-One and One-to-Two Correspondence with Geometric Illustrations' carried on this work, as did her 1918 paper 'Properties of a Certain Projectively Defined Two-Parameter Family of Curves on a General Surface'.

In 1917, Sperry began her career at UC Berkeley, where she would remain (with the one interruption in 1950–1952) until 1956, dividing her time between mathematics teaching and political causes, such as advocating for a ban on nuclear weapons testing, supporting the ACLU and League of Women Voters, and founding a school in Haiti. She was a lifelong Quaker, who served on the executive council of several Friends' organisations.

Olive Clio Hazlett (1890–1974)

Algebraist Olive Hazlett was the author of seventeen research papers, published over fifteen years, from 1915 to 1930. A graduate of Radcliffe and the University of Chicago, she spent seven years as an assistant professor at Mount Holyoke College, but was dissatisfied with the amount of time her teaching duties left her for research, and transferred to the University of Illinois, where she remained from 1925 to 1959. Her particular area of research was linear algebra and the study of functions on an algebra that are invariant under linear transformation. By 1930, her teaching load was so hefty that she essentially had to stop publishing original research, but even then, the stress of her schedule grew to be too much, and in 1936 she suffered a nervous breakdown, which took her two years to fully recover from, rejoining the University of Illinois in 1938. Her health remained precarious after her return, but that did not stop her from taking part as a member of the American Mathematical Society's secret Cryptanalysis Committee during the Second World War. By the war's end, her health had declined so precipitously that the university placed her on permanent disability leave.

Maria Pastori (1895–1975)

Born into a working-class Milanese family in 1895, Maria Pastori, along with her seven siblings, could not afford the private tutors and preparation schools that helped so many of her American and British contemporaries achieve mathematical renown. She relied extensively on earning scholarships through competitive examinations to claw her way to each new educational opportunity, beginning in the Milan public schools, then extending to an institution that provided the bare minimum of educational training for prospective teachers, allowing her to become an elementary school teacher. She and her sister Giuseppina, however, would wake up early and study as much mathematics as they could before Maria had to leave for her work, and by the age of 20 they were both able to take the state baccalaureate examinations, both earning high honours.

Pastori then took the Scuola Normale Superior entrance exam, again earning a scholarship through her high score, supplementing her scholarship income through side tutoring until she earned her PhD in 1920. Superior achievement on subsequent examinations earned her a place as assistant at the University of Milan, where she rose to become the chair of rational mechanics by 1939, and extraordinary professor by 1965. Her numerous publications focused on extending the differential calculus of Gregorio Ricci-Curbastro, otherwise known as tensor calculus, which Ricci-Curbastro had developed in the late nineteenth century with his student Tullio Levi-Civita, and which went on to become fundamental in the development of Einstein's theory of General Relativity. Pastori wrote about applications of tensor calculus, Einstein's relativity theory, and the mathematical principles underlying electromagnetism.

Sofia Aleksandrova Ianovskaia (1896–1966)

Unlike Tatyana Ehrenfest, who was a full 29 years of age when the 1905 Russian revolution occurred, and entering her forties when the Communist Revolution fundamentally remade Russian society, Sofia Ianovskaia represented a new generation of Russian women mathematicians whose youths were spent in revolutionary ferment, and whose adulthood was informed by those experiences.

She was born in Pruzhany in 1896, which was then part of Poland but is currently part of Belarus, but grew up in Odessa, where she attended the Odessa Higher Women's Courses. Aged 21 when the Communist Revolution broke out in 1917, she became an active Bolshevik in the subsequent Civil War, acting as a political commissar, in which capacity she was captured by anti-revolutionary White forces and scheduled for execution by firing squad.

That execution did not occur, and in 1924 she was able to continue her mathematical studies at the Institute of Red Professors in Moscow. Rubbing elbows with some of the great minds in Soviet mathematics, she was offered a teaching position at Moscow State University in 1926 before she completed her doctorate (which she received in 1935 from Moscow University). She began teaching courses in logic the year after receiving her PhD, which was a brave and controversial decision in the climate of Stalin's Soviet Union. Pure mathematical logic bore a reputation in Soviet circles as being an ideologically bourgeois field of study that lacked a firm basis in Marxist–Leninist principles. Ianovskaia risked much by lecturing in the field (though she protected herself to some degree in her writings by separating pure *mathematical* logic from the more ideologically tainted logic of Western *philosophy*), and more by extending her lectures on logic in 1946 to the philosophy department of Moscow State University.

Fortunately for her, in 1950 Stalin's shifting philosophical principles led him to compose a 'Letter on Marxism and Linguistics', which gave formal approval for the separation of linguistic studies from its ideological bourgeois roots. This publication, and her subsequent receiving of the Order of Lenin in 1951, allowed her to organise and chair a department of mathematical logic at Moscow State University, beginning in 1959, and in 1968 to write a full account of maths' place in the context of Marx's writings, which has since become useful to mathematicians in Communist regimes wishing to protect themselves from charges of ideological impurity.

Gertrude Blanch (1897–1996)

Born Gittel Kaimowitz in the politically Russian but historically Polish town of Kolno, Gertrude Blanch, the youngest of seven children, spent the first ten years of her life with her mother in Russia, waiting to be called to America by her father, Wolfe Kaimowitz. In 1907, she and her sister joined their mother in emigrating to the United States, settling in Brooklyn, New York, where she graduated from high school in 1914. That was also the year her father died, and in order to support the family and save up for her continued education, Blanch had to work in various clerical positions for the next fourteen years, finally taking college courses at Washington Square College in 1927 after the death of her mother.

In 1932, at the age of 35, Gertrude Blanch received her long-awaited bachelor's degree from New York University, changed her name from Kaimowitz to Blanch, and began attending Cornell University for graduate studies in algebraic geometry, receiving her PhD in 1935. This was not, however, a propitious time to be entering the work force, as the United States was just beginning to clamber its way out of the Great Depression on the strength of Franklin Roosevelt's New Deal programmes. Resorting to clerical work again, she was ultimately saved by FDR's Works Progress Administration program, popularly known as the WPA, where she found work in 1938 on the Mathematical Tables Project, a massive effort to design algorithms that compiled the values of desirable functions in an age before the existence of electronic calculators.

Blanch's job involved selecting which functions would be most useful to know the values of, and designing algorithms to efficiently calculate those values that could be handed off to her team of 450 human calculators. During the Second World War, in addition to its role preparing standard computations of transcendental functions, the MTP also performed calculations central to army operations ranging from the Manhattan Project to the storming of the beaches at Normandy. After the war, Blanch worked a variety of numerical analysis positions until coming to rest at last as a senior mathematician studying problems of air flow and supersonic flight at Wright-Patterson Air Force Base in Ohio.

Marguerite Lehr (1898–1987)

Like Anna Johnson Pell Wheeler, Marguerite Lehr had to work from humble beginnings to earn her higher education by means of her superior mathematical gifts. She was born in Baltimore, Maryland, to Margaret Kreuter and George Lehr, a humble grocer. Well, he was a grocer – his humility is not readily attested to one way or the other in the historical records. The eldest of five children, she was the only one to attend college, graduating from Goucher College in 1919, having taken classes there from Clara Latimer Bacon, whom we met earlier.

In 1923, she was able to study at the University of Rome under an American Association of University Women fellowship, and began attending Bryn Mawr upon her return in 1925, where both Charlotte Angas Scott (totally deaf at that point in her career) and Anna Johnson Pell Wheeler were teaching. She earned her PhD from Bryn Mawr in 1925 for her work on plane quintic curves (i.e. curves of degree 5) possessing five cusps. Lehr continued her association with Bryn Mawr for the next four decades, teaching there from 1924 to her retirement in 1967.

Lehr is also notable as one of the earliest professors attempting to use the medium of television to create educational mathematics content, hosting the programme 'Invitation to Mathematics' from 1953 to 1954 (which, if you can find the original clips from somehow, please let me know!), and working as a consultant for NBC on a mathematical programme they ran in 1957. One of her last publications was a 1971 appreciation of the life and work of Charlotte Scott.

Gertrude Mary Cox (1900–1978)

Cox is best known as a founder of the Biometric Society (1947), and early advocate of the use of experimental statistics in psychological and biological studies. Born in Dayton, Iowa, she received her bachelor's and master's degrees from Iowa State College in Ames, Iowa, in 1929, and 1931, after having given up on plans to become a deaconess. She earned her master's from a statistical study of teacher effectiveness as measured by subsequent student success.

Her career was mainly spent at the University of North Carolina, and North Carolina State University, where she was the head of the department of experimental statistics, and director of the Institute of Statistics until her retirement from NCSU in 1960. During that time, she and statistician W.G.

Cochran wrote and published *Experimental Designs* (1950), which stood for many years as the bible of experimental statistics design methodology. After her retirement from NCSU, she worked for five years as head of the Statistics Research Division at North Carolina's Research Triangle Institute, and subsequently undertook the presidency of the Biometric Society in 1968–9.

Nina Karlovna Bari (1901–1961)

One of the world's undisputed masters of trigonometric series, Nina Karlovna Bari published fifty-five research papers and a 900-page monograph over the course of her thirty-eight years of research, representing a truly remarkable pace of rigorous and foundational work even without taking into consideration the fact that the destruction and displacements of the Second World War lay directly at the heart of that time. Born in Moscow, she attended LO Vyazemska's High School for Girls, at the end of which she opted to show her mathematical muscle by taking the boys' final exam rather than the generally less rigorous girls' one.

In 1918, Moscow State University was opened to women enrolment for the first time by Communist decree, and Bari was the first woman to join the physics and mathematics department, falling quickly into the orbit of Nikolai Luzin, who studied function theory and advocated the constructive method of proof, whereby the existence of objects was proven by providing techniques for constructing them. Bari was particularly gifted in the creation of constructive proofs, an ability which underlay her extraordinary insights into trigonometric series.

In 1921, she became the first woman to graduate from Moscow State University, and in 1922 the first woman to address the Moscow Mathematical Society, presenting her results about how to determine the uniqueness of a particular trigonometric expansion of a given function. The problem of whether a given trigonometric representation of a function (say, a Fourier Series) is the *only* possible trigonometric representation of it, has its roots in a problem Heine proposed to a young Georg Cantor in 1869. Bari's work sought the conditions for uniqueness, and she marshalled her startling abilities in constructive proof to achieve that end, winning the Glavnauk Prize (as well as her PhD) for her 1925 paper on series uniqueness.

Over the thirty-eight years of her academic career (terminated in 1961 when Bari either intentionally stepped, or accidentally fell, in front of an oncoming train), she produced articles spanning not only her area of expertise in trigonometric series, but pieces on the superpositions of absolutely continuous functions, biorthogonal systems, and the approximation of functions, in addition to her two-volume monograph on Trigonometric Series that was published in the year of her death and became *the* book of reference on the subject for generations to come.

Edna Ernestine Kramer Lassar (1902–1984)

Born in Manhattan to Jewish immigrants with strong educational beliefs, Edna Lassar and her two siblings were all academic over-achievers, but Edna's skills in mathematics were particularly noticed by her teachers in Wadleigh High School and fostered. She attended Hunter College for her undergraduate work, and earned her master's at Columbia University in 1925, and her PhD there in 1930 for her work on polygenic functions of dual variables (polygenic functions are those that possess an infinite number of derivatives at a point, while dual variables are variables that have some form of mutual dependence, in Lassar's case she considered variables w such that w was itself of the form u + jv).

After receiving her PhD, Lassar turned from pure mathematics and became more interested in writing about the history of, and methods of teaching, mathematics. She married Benedict Lassar, a French teacher, in 1935, shortly after stepping down from her job as an instructor at Montclair College and resuming what she thought was the more Depression-proof career of teaching at Thomas Jefferson High School, in Brooklyn. In 1948, she also began teaching at Brooklyn Polytechnic, a position she maintained until 1965. By the mid-1970s, the effects of Parkinson's disease began to affect the quality of Lassar's life, and in 1984 she died of pneumonia at home.

Mina Spiegel Rees (1902–1997)

Like her contemporary, Edna Lassar, Mina Rees was a product of Hunter College (where she graduated in 1923), and Columbia University (where she earned her master's the same year as Lassar, in 1925). She taught at Hunter High School in the 1920s, leaving in 1929 to study at the University of Chicago, earning her PhD in 1932 on the topic of Division Algebras. After teaching at Hunter College from 1932 to 1940, and working for the Office of Scientific and Research Development in the Second World War, a rare opportunity arose after the war to become head of the mathematics branch at the Office of Naval Research, where she spearheaded studies of mathematical algorithm development for computer programs, and the development of new generations of computers to handle those algorithms. She advocated for the implementation of several computing features that became standard in the industry, including the use of transistors over vacuum tubes, magnetic core memory, multiple inputs and visual displays.

Rees worked at the ONR from 1945 to 1953, laying down policies that later influenced those of the National Science Foundation when it was established in 1950. In 1953, she returned to Hunter College as dean of faculty, serving in that position until 1961, when she became the first dean of graduate studies at the City University of New York, which was founded in that year (though its constituent colleges had been in existence since the mid-nineteenth century). She was honoured in 1989 by the IEEE Computer Society with their Computer Pioneer Award.

Ludmila Vsevolodovna Keldysh (1904–1976)

Ludmila Keldysh produced a remarkable amount of work in set theory, topology and geometric topology during a life often deeply fragmented by the tragedies of war and totalitarian persecution. Born in Orenburg, Russia, on 12 March 1904, she was the oldest of seven siblings in a family whose ancestors came from the nobility and featured a number of Russian generals, facts that would work against her in the post-1917 Soviet landscape. By 1909, her family had moved to Riga, where her father lectured at the Polytechnic Institute, only to have to flee the city with the German invasion of Latvia in 1915.

Evacuated to Moscow in much-reduced circumstances, the large family had to make do as best it could with uneven access to food and money. Nevertheless, Keldysh managed to attend high school in Ivanovo-Voznesensk, where a guest lecture by Nikolai Luzin (whom we met above in the tale of Nina Bari) inspired her to take up the study of mathematics. She entered the Physics and Mathematics Department at Moscow State University, and by 1923 had joined Luzin's research seminar. She graduated in 1925, and for the next five years information on her life is sketchy until she emerges in 1930, publishing her first mathematics paper, and taking up work at the Moscow Aviation Institute. That paper was on a topic that she would return to repeatedly: Borel sets and their classification (a Borel set is a set in a topological space that is made up of the countable union, countable intersection, or relative complements, of open sets in that space).

After 1934, Keldysh worked at the Mathematical Institute of the Soviet Academy of Sciences, during which time her uncle was murdered, and parents were arrested but released, in the great Stalinist purges that stretched across the early 1930s. Subsequently, with the German invasion of the Soviet Union in 1941, she and the rest of the faculty of the Institute had to flee Moscow, but when they arrived in Kazan, instead of being given rooms as evacuees, like most members of the Institute, she and her children (from her marriage to Pyotr Sergeyevich Novikov in 1935) were classed as mere refugees, and stuffed in the local gymnasium with others of that classification, without warm winter clothes or personal possessions or any reliable means of attaining sufficient food. They lived in this condition for a month, until Novikov arrived and the family was given a cold room in a dormitory. For two years, the family survived on a mixture of coarse rye flour and water known as zatirka.

After the war, Keldysh turned increasingly toward geometric topology, and particularly to the subject of embeddings, which investigates the preservation of the properties of an object when that object is considered as a part of a larger or different space. For example, if you look at a set that is a field, like the Rational numbers under addition and multiplication, you can say that it is embedded in the field of the real numbers, because the addition and multiplication operations that hold for real numbers, when restricted to the Rationals, work the same as the addition and multiplication of the original set of Rationals. That idea can be expanded to topological embeddings, which investigates how things like the open sets of the relevant spaces are preserved, and these sorts of investigations formed much of Keldysh's work in the 1960s, and which form the subject of her 1966 monograph *Topological Imbeddings in Euclidean Space*.

Sophie Piccard (1904–1990)

Sophie Piccard's parents were both professional academics, her father a Swiss university professor, and her mother, Eulalie Güée, a French historian who wrote a five-volume history of the Russian Revolution. She was born in St Petersburg, but the family fled Russia, emigrating to Switzerland in 1925 after the Russian Civil War (1917–1923) resulted in the victory of the Communists. In Switzerland, Piccard studied at the University of Lausanne, where she received her scientific licence in 1927, and her PhD in 1928 in the field of probability.

The early death of her father compelled both Sophie and her mother to take whatever work they could to make ends meet, her mother taking a job as a dressmaker, and Sophie as an actuary and administrative secretary until 1936, when she was able to take an academic position again, as assistant in geometry at the University of Neuchatel, then professor of geometry in 1938, and gaining the chair of higher geometry and probability theory in 1943.

In 1939, she published *Sur Les Ensembles de Distances des Ensembles de Points d'un Espace Euclidien*, which investigated the sets of distances that arise from different sets of points in Euclidean space, and developed congruence conditions for Golomb rulers. She followed this in 1942 with the book *Sur les Ensembles Parfaits*, about perfect sets, which are closed sets that contain no isolated points (which are what they sound like, points around which you can draw a neighbourhood that contain no other points in the set).

Sophie Piccard died in Switzerland in 1990, at the age of 85.

Maria dei Conti Cinquini-Cibrario (1905–1992)

The author of nearly 100 mathematical articles, Maria Cinquini-Cibrario received her PhD from the University of Turin in 1927, and did postgraduate work at the Scuola Normale Superiore in Pisa. Her research included the classification of mixed type, second order partial differential equations, the uniqueness of solutions to systems of partial differential equations, and systems of non-linear hyperbolic equations.

Käte Fenchel (née Sperling) (1905–1983)

Käte Sperling was born in Berlin on 21 December 1905 to Rusza Angress and Otto Sperling. Her father, Otto, left the family when she was young, creating deep financial hardships that Käte had to overcome in order to obtain a thorough education. She was able to earn a scholarship to attend a private high school, and the father of one of her friends was generous enough to underwrite her studies at the University of Berlin. She was asked by the mathematics department to write a thesis, but her sense of pragmatism told her that academic jobs for women were a rarity, and that she would be better served switching her concentration to training as a teacher.

She served as a high school teacher from 1931 to 1933, when Nazi race laws precluded her, as a Jew, from continuing in that capacity. Thereupon, she and fellow student Werner Fenschel emigrated to Denmark in 1933, where they married in December. In Denmark, she took work as a part-time secretary to a mathematics professor, in which capacity she was able to continue her mathematics studies and aid other Jewish refugees in finding their way out of Germany. Her first paper was published in 1935 on groups of linear transformations, followed by another in 1937 on the topic of everywhere dense vectormodules (modules are essentially vector spaces that pull their scalars from a ring instead of a field).

When Germany invaded Denmark, the Fenchels had to flee once more, this time to Sweden, where they remained for the duration of the war. Her first mathematical publication after the war did not occur until 1962, when she wrote about odd-ordered groups, and again about the structures of finite groups. Her last paper, on one of Frobenius's theorems, was published in 1978, five years before her death.

Ruth Moufang (1905–1977)

The namesake of Moufang planes and Moufang loops, Ruth Moufang was a foundational figure in the algebraic analysis of projective planes. Born in Frankfurt am Main, she and her artistically inclined older sister Erica attended the Realgymnasium in Bad Kreuznach, where they helped their mathematics teacher, Wilhelm Schwan, by doing the illustrations for his mathematics textbook. Ruth passed her Abitur in 1924, and began attending the Johann Wolfgang von Goethe University in Frankfurt am Main, studying for her teaching degree, which she received in 1929, while carrying on mathematical investigations that would lead to her PhD in 1931. Her dissertation involved the reworking of David Hilbert's famous plane geometry axioms into projective geometry formulations.

In a burst of intellectual energy, from 1931 to 1934 Moufang produced the seven papers that laid down the basic principles of the algebraic analysis of projective planes, work that in the 1950s would lead to the development of 'Moufang planes' (projective planes in which the little Desargues theorem holds) and 'Moufang loops' (group-like structures in which the Maclev algebra holds, i.e. $xy = -yx$). On the strength of this fundamental mathematical work, Moufang received her Habilitation in 1937, which should have set her on the path of an academic career, but she was denied a professorial position by the Nazi hierarchy, which declared that a woman could not exert proper 'leadership' over a class that is predominantly male. She would not become a full professor until well after the war, when in 1957 she was the first woman to be named a full professor at the University of Frankfurt.

Rozsa Peter (née Politzer) (1905–1977)

Born Rosa Politzer to a Hungarian Jewish family, Rozsa Peter (who changed her name to a less-German, more-Hungarian variant in the 1930s), grew up in a Hungary that was in a state of seemingly perpetual turmoil. Food shortages during the First World War gave way to dissolution, Communist revolution and White counter-revolution after the war. Peter's father was a lawyer and was able to pay for her to attend a decent girls' school, from which she graduated in 1922. She subsequently attended Pazmany Peter University (now Eotvos Lorand University), graduating in 1927. Her father had desired that she study chemistry, but she caught the mathematics bug while attending the lectures of Lipot Fejer, and subsequently switched her field of study. While at university, a fellow student, Laszlo Kalmar, sparked her interest in recursive functions, which would form the subject of her work in the 1930s.

Following her graduation, Peter was without steady employment for eighteen years, picking up whatever odd jobs she could during the Depression as a mathematics tutor, and journal editor, while continuing her private studies. She received her PhD in 1935, but the rise of the fascist Bela Imredy to the position of prime minister in 1938 put academic employment even further out of her reach when his government proposed a harsh Second Jewish Law, put into effect by his successor in 1939, which pushed Jews out of the teaching professions entirely.

During the 1930s, Peter worked primarily on recursive function theory, a branch of mathematics and logic that had its origins in a 1923 paper of Thoralf Skolem, and to which she would contribute foundationally by investigating the phenomenon of higher order recursions. In 1976, she wrote an important book highlighting the importance of recursion theory to computer theory, *Rekursive Funktionen in der Komputer-Theorie*, which appeared in English translation in 1981.

In 1945, she finally found long-term employment, at the Pedagogiai Forskola, a teacher's college, and later at Eotvos Lorand University. She retired in 1975, two years before her death at the hands of cancer. If you want to get a good feel for her, both as a mathematician, and a human being, try and flag down a copy of her 1945 book *Playing with Infinity*. It is a hoot.

Winifred Sargent (1905–1979)

Winifred Sargent was a mathematician's mathematician, devoted entirely to her craft, with little interest in the conferences and seminars that served as opportunities for social and professional advancement for so many of her colleagues. Born to a Quaker family, she had a solid work ethic that admitted of no distractions, and no interest in the pursuit of merely personal glory. Her early education was funded by a series of scholarships she had earned, culminating in a Derby scholarship, Mary Ewart scholarship, and state scholarship that allowed her to attend Newnham College, Cambridge, beginning in 1924. She was awarded a first-class degree in 1927, but left the school in 1928 when she failed to meet her own perfectionist standards for her research, teaching at Bolton High School for three years.

The itch to carry out more substantive research soon reasserted itself, however, and in 1931 she took up a position at Westfield College, and began her studies of Lebesgue integration and Cesaro summation (both of which provide ways of thinking about how integration might work for functions that at first glimpse seem to defy that process). By the 1950s, she was investigating the equally fascinating phenomenon of fractional integration and differentiation, which is an area of mathematics interested in answering questions like, 'What precisely does it mean to take the 1.7th derivative of a function?' Her work in the 1960s centred around BK spaces, which are sequence spaces (spaces whose elements are infinite sequences of numbers) equipped with a norm (a way to measure the distance between elements and length of elements) that allows them to be considered as Banach spaces.

Sheila Macintyre (née Scott) (1910–1960)

Sheila Scott was born in Edinburgh, where she received most of her early education, graduating from the Edinburgh Ladies' College in 1926, and receiving her master's at the University of Edinburgh in 1932. Her professors encouraged her to obtain a second bachelor's degree to deepen the more generalised mathematical knowledge she had accrued in Edinburgh, and so she began attending Girton College, receiving another undergraduate degree in 1934. At this time, she decided that research did not hold the interest for her she had hoped it would, and she took up a string of teaching posts until her marriage to Archibald James Macintyre in 1940.

Macintyre had a teaching position at the University of Aberdeen, and upon their marriage Scott was also invited to lecture there. With the advent of the Second World War, she taught more classes at Aberdeen to fill in for the male professors who had gone off to war, and taught special courses for the War Office and Air Ministry. In 1947, she completed her doctoral thesis at the University of Aberdeen, 'Some Problems of Interpolatory Function Theory'.

Her research had to be carried out in and around the care of her children, born in 1944, 1947, and 1950, but during the period 1947–1958 she still managed the publication of ten articles, mostly on the theory of complex variable functions, particularly questions about the convergence of such functions. One of her more notable papers included a determination for the range of possible values of the Whittaker Constant. Previous researchers had determined that W must be less than .7399, while Macintyre's 1947 paper managed to improve that estimate, putting the upper limit at .7378.

In 1958, Macintyre's husband was given a position at the University of Cincinnati, where she became a visiting professor briefly, before dying of cancer in 1960.

Marjorie Lee Browne (1914–1979)

Marjorie Lee Browne was the third African American woman in history to receive a mathematics PhD (Euphemia Haynes was the first, in 1943, and Evelyn Boyd Granville the second, in 1949). Born in Memphis, Tennessee, her mother died when she was 2 years old, leaving her to be raised by her father and eventual stepmother. A precocious child, she demonstrated abilities in multiple areas, including tennis (winning the Memphis City Women's Championship in 1929), music, and of course, mathematics.

She attended Howard University, a historically black college in Washington DC that had been established in 1867 during Reconstruction, where she graduated *cum laude* in mathematics in 1935. Receiving a teaching fellowship at the University of Michigan allowed her to support herself as she worked on her doctorate thesis, which was completed in 1949 on the topic of one-parameter subgroups of topological and matrix groups. After completing her thesis, Brown found a position at North Carolina Central University as a professor and researcher, where she remained from 1951 to 1970, carrying on her studies in combinatorial topology, differential topology and the use of computing to improve numerical analysis. In 1960, she wrote the grant application that brought an IBM computer to NCCU, arguably the first instance of a computer being made available to the researchers and students of a historically black college.

Dorothy Lewis Bernstein (1914–1988)

Dorothy Lewis Bernstein was the child of Jewish Russian immigrant parents who had resolved to send their children to college, even remortgaging their home in order to scrape together enough money to do so. Bernstein was born in Chicago, Illinois, but received her education in Wisconsin, first at North Division High School, and later at the University of Wisconsin, where her intention to major in journalism was derailed by the encouragement of Professor Mark Ingraham to try out mathematics. She received her bachelor's and master's degrees simultaneously in 1934. In 1939, she received her doctorate after an eight-hour oral examination that was four times the length of those given to the male students from more established East Coast universities.

At Brown University, she was actively discouraged from teaching male students, and when she asked about where she should look for a teaching job, she was told to look nowhere east of the Mississippi, because they wouldn't hire a woman, and nowhere in the South, because they wouldn't hire a Jew. She ultimately found a position at Mount Holyoke, where she taught from 1937 to 1940, then at the University of Rochester, a post she held from 1943 to 1959, and ultimately at Goucher College (where Clara Bacon had been a trailblazing instructor, decades prior) from 1959 to 1979, when she retired.

Bernstein was interested in the application of partial differential equation existence theorems to problems in computing. After the Second World War, the Office of Naval Research (where you'll remember Mina Rees was a driving force) was investigating how to calculate the solutions to complex partial differential equations. Part of doing that is knowing under what boundary conditions, and in what locations, a given system of equations is likely to have valid solutions, which was Bernstein's field of expertise. In 1950, she published *Existence Theorems in Partial Differential Equations*, which served as an important reference book for the ONR and other advanced computing projects. In line with these interests, Bernstein was also instrumental in bringing an IBM 1620 computer to Goucher, serving as director of the resulting Goucher Computer Centre for six years.

Bernstein was elected president of the Mathematical Association of America in 1979, the first woman to ever hold that position.

Ruth Bari (née Aaronson) (1917–2005)

No relation to Nina Karlovna Bari, whom we met above, Ruth Aaronson was an American mathematician born in Brooklyn, where she went on to earn her bachelor's degree in 1939 from Brooklyn College, and her master's from Johns Hopkins in 1943. She had originally intended to stay at Johns Hopkins until she received her doctorate, but with the conclusion of the Second World War the university strongly suggested that she should give up her position so that a returning soldier could fill it, and she did so. She then married Arthur Bari and for two decades raised a family, not returning to mathematics until 1966, when she re-entered Johns Hopkins and received her doctorate at the age of 48.

Bari's areas of study were graph theory, and the chromatic polynomials arising from different graphs (here we are thinking of a graph as a collection of vertices and edges. A proper 'colouring' of such a graph, is a graph wherein each vertex is assigned one of a limited number of colours, and no two vertices connected by an edge have the same colour. The chromatic polynomial is a polynomial that tells you, for a given graph, and a given number of possible colours, the number of unique proper colourings that exist.)

After receiving her doctorate, she taught at George Washington University until being compelled by law to retire at age 70.

Anne Phillipa Cobbe (1920–1971)

Born in Bedfordshire to a family of prominent Irish ancestry, Anne Cobbe had a youth punctuated by loss. Her father died when she was 10, followed by a brother who was killed in the Battle of Britain when she was 20. She attended Somerville College at Oxford from 1939 to 1942, and then placed her mathematical talents at the service of the Royal Navy, working on the development of better ways of arriving at decisions for complicated real-world problems.

Returning to Oxford after the war, she received her master's degree in 1946 and began research at Lady Margaret Hall, where she worked on homological algebra, a field that had grown into its own arena of research in the 1940s after a century of being an understudied outgrowth of topology. She earned her PhD in 1952, then spent some time as a lecturer at Lady Margaret Hall, but opted in the end to become a tutor at Somerville, preferring the intellectual exchanges that could be had in one-on-one sessions to those generated in a lecture hall.

Her work in the 1950s centred on cochain complexes and cohomology groups. Her health precipitously declined in 1969, and she died two years later, at age 51.

Evgeniia Aleksandrovna Bredikhina (1922–1974)

Evgeniia Bredikhina is a mathematician whose presence grows a little dimmer each year in the collective memory of mathematical historians. She published thirty-five papers exclusively on the subject of almost periodic functions (which are what you probably are thinking they are), and was decorated by the Soviet Union multiple times for her work both as a teacher and researcher. In 1939, she began attending the University of Kazan, but transferred to the University of Leningrad in 1940. During the Siege of Leningrad in 1941, she worked in a hospital tending to the wounded until 1942, when she was evacuated from the city to Kuibyshev, where she finally completed her undergraduate degree in 1945. She received her PhD in 1955 from the University of Leningrad, and her DSc from the University of Lvov (now known as Lviv) in 1972, two years before her death.

Cathleen Morawetz (née Synge) (1923–2017)

A recipient of the National Medal of Science in 1998, Cathleen Synge hailed from a scientific family. Her father was a mathematician interested in the geometries involved in general relativity, her mother had been a student of mathematics for a number of years, and her uncle invented the near-field scanning optical microscope. She graduated from the University of Toronto in 1945, and earned her master's at MIT in 1946 and her PhD at New York University in 1951. After some time spent at MIT, she took a purely research position at the Courant Institute of Mathematical Sciences at NYU where she began her investigations into the differential equations that model fluid movement.

On 2 October 1947, the first transonic flight had occurred, and interest in transonic fluid flow was at a fever pitch. Morawetz's work in the 1950s posited the impossibility of shockless transonic air foils by showing that any continuous air flow around a surface would produce shocks as soon as small perturbations in its shape inevitably occurred, and developed new wave equations for transonic fluid flow around an object. In the 1960s, she continued her studies of fluid flow, but added to them studies on the scattering of electromagnetic waves striking a surface. Over the course of the 1980s, 1990s, and early 2000s, she received more awards and honours than it is remotely seemly to try and record her, though they were certainly no more than her rich career deserved.

Carol Ruth Karp (née Vander Velde) (1926–1972)

The youth of Carol Ruth Vander Velde resembled nothing so much as the lives of the teenagers of the film *Footloose* before Kevin Bacon came to town. Her religiously strict Dutch parents forbade dancing, movie-going, and disapproved sternly of card-playing, so that the young Carol's main outlet for youthful rebellion was to be found in her playing of the viola. Born in Forest Grove, Michigan, her family moved to Alliance, Ohio, then to Bremen, Indiana, and then moved again, to Ionia, a pattern of rootlessness that Vander Velde perpetuated in the early part of her academic career.

She received her undergraduate degree from Manchester College in Indiana, then her master's in 1950 from Michigan State College. From 1951 to the awarding of her PhD in 1959, Vander Velde (who became Carol Karp upon marrying Arthur L. Karp in 1951), while ostensibly a graduate student at the University of Southern California, actually spent that decade in a series of one- to two-year teaching positions, from New Mexico to Berkeley to Japan, ultimately ending up at the University of Maryland in 1958, where she finally settled down, and remained until her death in 1972.

Karp made the University of Maryland into a research hub for logicians, arranging positions not only for a number of young and established talents but also attracting the world's most renowned logicians to the Maryland Mathematics Colloquium and Logic Seminar. Her own research centred around the development of a new field of logic, known as infinitary logic, in which the use of infinitely long statements or infinitely long proofs is allowed. In 1956, she was the first to show how to extend first-order logic (what we think of as regular mathematical logic) to statements employing infinite unions and intersections. Karp's 1959 dissertation was on 'Languages with Expressions of Infinite Length', which she expanded into a full-length book in 1964 of the same title that established many of the ground rules for the discipline. Karp demonstrated how certain groups and fields, which could not be well characterised under finite logical structures, could be using infinitary logic.

In 1969, Karp was diagnosed with cancer, but continued teaching as long as she was physically able, and even attempted to produce a new book on recent developments in infinitary logic in the time remaining to her, a project which her declining health did not allow her to complete before her death in 1972.

Joan Sylvia Lyttle Birman (b. 1927)

For every secondary school student freaking out because they don't have every aspect of their future nailed down by the age of 17, Joan Birman is the great example of the virtue of chilling out and letting your life's direction unfold in the fullness of time. When she entered Swarthmore College, she was undecided between a career in physics, biology, or astronomy. She knew that she liked living in cities, and that astronomers in that era tended to have to live in remote locations, so she eventually eliminated that as her potential field of study. She also eliminated Swarthmore as the place she wanted to spend her undergraduate career, and transferred to Barnard College, which is another good lesson for students – if you don't like where you are, you *are* allowed to leave!

At Barnard she majored in mathematics, receiving her degree in 1948, but then experienced a crisis in confidence in her mathematical abilities, and instead of continuing on to graduate studies, decided to work for an engineering company until that job became too rote and tedious. One of her physics teachers at Barnard offered her a position as an assistant, which would allow her to carry on with her master's degree, and having learned a little about herself and the types of jobs offered to people in the private sector who didn't have advanced degrees, she readily agreed, receiving her master's in physics in 1950, the same year she married physicist Joseph Birman. For five years she worked as a systems analyst, and over the subsequent five years, she had three children and had to give up working. During that half decade, she came to miss the world of mathematics that she had enjoyed as a youth, discussing Euclidean proof on the phone with her friends for hours, and as soon as her last child was born, she signed on with New York University's Courant Institute to receive her master's in mathematics in 1961.

She had some difficulty in obtaining an advisor for her PhD, as various professors expressed scepticism about her age and diverse intellectual background; ultimately, she was taken in by Wilhelm Magnus, who had a habit of taking in what he called 'strays' – PhD candidates who for some reason or other had been turned away by his colleagues and needed a warm and understanding place to carry out their research. He turned Birman, who expressed an interest in pure mathematics and especially topology, on to braid theory, which is one of those branches of mathematics that you cannot help but fall a little bit in love with as soon as you see it. Essentially, it investigates the different ways of connecting two sets of points arranged in parallel rows, allowing you to weave the connecting strands over and under each other in patterns that look like braids.

Birman received her PhD in 1968 for her work on braids, generalising the concept of braid groups, which had been typically applied to the Euclidean plane, to arbitrary manifolds. Seven years later, in 1975, she published the classic text *Braids, Links, and Mapping Class Groups*. Birman taught at the Stevens Institute of Technology from 1969 to 1973, and then at Barnard College, from 1973 until her retirement in 2004. Her work on braids led to studies of knot theory, operator algebras, mapping class groups, and contact topology. She continued publishing papers into the twenty-first century, including an article on distance algorithms published in 2016 as she approached her ninetieth birthday.

Marian Pour-El (née Boykan) (1928–2009)

Over the last half century of humanity's relationship with computers, we have grown accustomed to the idea that, if we want a value for a function or the solution to a well-stated problem computed, all we need to do is go to our nearest computer and ask it do so, and then, somehow, an answer will happen. Certainly, this is true for the things most of us are called upon to compute on any given day, but do there exist problems that, even with all our vaunted computing power, we still cannot solve?

Questions such as this are the domain of computability, or recursion, theory, which has been around since Alan Turing's work in the 1930s to define what sorts of things can and cannot effectively be calculated by machines. Among the most interesting of computability theorists in the middle to late years of the twentieth century was Marian Boykan, born on 29 April 1928, in New York City. Had she been a boy, her gifts in science and maths would have funnelled her towards the Bronx High School of Science, founded in 1938 and widely known as the secondary school that has produced the most Nobel laureates in the world. It was, however, at the time exclusively a boys' school, and so she had to attend Hunter College High School instead, and then move on to Hunter College because her parents refused to pay the higher private university tuitions.

She graduated from Hunter in 1949, and then attended Harvard for her graduate work, where she was the only woman in the department, often shunned by her male classmates, and lacking basic restroom facilities in the building where she studied. It was a psychologically gruelling time, as Boykan explained to Claudia Henrion for the portrait of her life and work in Henrion's 1997 book, *Women in Mathematics: The Addition of Difference*. Amenities for basic comfort were denied to her either out of malice or

163

simple institutional inertia, and she lacked both mentors of her own gender, and colleagues working in her preferred field of mathematics, which was logic.

After she received her PhD from Harvard in 1958, Boykan (now Pour-El after her marriage to biochemist Akiva Pour-El) taught primarily at two universities, Pennsylvania State University until 1964, and the University of Minnesota from 1964 to 2000 with one year off as a visiting professor at the University of Bristol. Her speciality, as it developed over the 1960s and 1970s, was the area of computability theory, whereby she found not only special unsolvable cases of problems that were long assumed to be eminently solvable, but investigated the sort of problems that can be computed by analogue, as opposed to digital, computers, by harnessing the continuous nature of their inputs. The result of that investigation is contained in the Shannon-Pour-El Thesis.

In 1989, she collaborated with J. Ian Richards on the book *Computability Analysis and Physics*.

Mary Catherine Weiss (née Bishop) (1930–1966)

In her thirty-five brief years of life, Mary Weiss wrote twenty mathematical papers on diverse subjects. She was born in Wichita, Kansas on 11 December 1930, but the family moved to Chicago, where she would spend most of her life, after she graduated from grade school. Her father, Colonel Albert Bishop, who enjoyed mathematics as a hobby, had died just before she was 2 years old, and the books he left behind sparked the interest of first Mary's brother Everett, and then her.

She attended the University of Chicago for her entire undergraduate and graduate career, receiving her PhD in 1957 for her dissertation 'The Law of the Iterative Logarithm for Lacunary Series and Applications to Hardy-Littlewood Series', which dealt with a topic she would often return to in her brief career, lacunary series. These are series that cannot be analytically continued outside of their original region of convergence, which is to say that, if your series, represented by the function $f(x)$, converges in a region A, there is no way to make a 'bigger' analytic function $F(x)$, that equals $f(x)$ everywhere in A, but has as its domain a region larger than A. Weiss's work centred on elucidating the connection between lacunary trigonometric series and the series of random independent variables that are central to probability theory.

Mary Weiss also produced work on differentiability conditions, singular integrals and Hardy spaces, often in collaboration with her husband, Guido

Weiss (1928–2021). She died suddenly, three weeks after assuming a position on the faculty at the University of Illinois in Chicago.

Mary Warner (née Wynne) (1932–1998)

Mathematics has the virtue of being the most portable of the STEM fields, something you can do to at least some degree virtually anywhere, making careers like that of Mary Warner, whose life was constantly upended by the job requirements of her husband, possible. In 1956, she married Gerald Warner, who was in the Diplomatic Service, and over the subsequent two decades she performed her mathematical researches wherever Gerald's job took him, from China to Burma to Poland to Malaysia, which delayed her obtaining of a PhD until 1968, fifteen years after she had started her graduate studies.

Born in Carmarthen, Wales, to a father who had some fame as a prominent secondary school headmaster, Warner had ready educational support at hand as she developed her mathematical and scientific gifts from a young age. She won a scholarship to attend Somerville College, Oxford (one of two women's colleges opened at Oxford for women in 1879, women not being admitted to the men's colleges until 1974), beginning in 1951, and graduating in 1953. Her graduate career began at Oxford under the supervision of homotopy theorist (and nephew of Alfred North Whitehead) John Henry Whitehead. He had built up an algebraic topology group at Oxford, which Warner joined, producing her first paper in 1956.

While she was in Oxford, Gerald Warner entered her life and, upon finding out that he was due to serve in China, the couple married, and Mary went with him, continuing her studies with mathematical colleagues there who were growing ever more nervous about their position as intellectuals in Mao's China. The Warners returned to England in 1958, only to leave for Burma shortly after the birth of their second child in 1959, setting a pattern of movement that was the status quo for the Warners until 1976, when Gerald's field service was concluded and Warner could settle down at City University for the work that would make her fame: developing the field of fuzzy mathematics.

Fuzzy mathematics had its origin in 1965 with Lotfi Zadeh's book *Fuzzy Sets*, and has a wonderful idea at its core – What if, instead of always answering Yes or No when asked if an item is part of a mathematical set, we could answer 'Kind of'? The fuzzification of maths involves working out the consequences of the idea of partial inclusion – what happens to beloved mathematical principles if we allow elements to be rated on a scale of, say, 0 to 1 for how much they belong to a given set? Warner's work in the late 1970s and 1980s focused on extending fuzzy ideas into the realm of topology.

Hoàng Xuân Sinh (b. 1933)

The first Vietnamese woman to earn a PhD in mathematics, Hoàng Xuân Sinh was also the first woman to become a full professor in Vietnam in any scientific field. She was born on 8 September 1933, in one of Hanoi's rural districts, as one of seven siblings in the home of a fabric merchant. She completed her bachelor's in French and English in Hanoi in 1951 in the midst of the First Indochina War (1946–1954) and then travelled to Paris to take up another undergraduate degree in mathematics (Vietnam was still considered a part of the Indochinese Federation until 1954 and it was not uncommon for Vietnamese students to choose France as the location for their education abroad).

Hoàng remained in France, studying at the University of Toulouse until 1959, when she returned to what was then North Vietnam to take up a position as a professor at the Hanoi National University of Education. The war with the United States had its impact on opportunities for mathematical exchange, particularly on the frequency of visiting lecturers and professors from outside of Vietnam, so when Alexander Grothandieck (whose name you might remember from Raman Parimala's story), known for his foundational work in modern algebraic geometry and his pacifist political stance, came to Hanoi to teach in 1967, Hoàng was eager to hear about the newest developments in mathematics.

Grothandieck and Hoàng became colleagues, and when he returned to France, he adopted Hoàng as a PhD correspondence student, allowing her to remain in North Vietnam while continuing her work towards a doctorate, which she earned in 1975 for her thesis on Gr-categories, which proved useful in the development of 2-group theory in the early 2000s. In 1988, she founded, and became president of, Vietnam's first private university, Thang Long University.

Vivienne Malone-Mayes (1932–1995)

Upon entering Fisk University in 1948 at the age of 16, Vivienne Malone's ambition was to earn a degree in medicine – until she met her future husband, James Mayes, who was studying dentistry and thought that a marriage wherein two people had similar professions was doomed to fail, so encouraged her to switch to mathematics. Fortunately, Fisk University at the time had Evelyn Boyd Granville on its faculty (whom we met above as the second African American woman to earn a mathematics PhD in the US), who was able to take Malone under her wing, and mentor her on what life would be like as a black woman in a field dominated by Caucasian men.

Malone earned her bachelor's in 1952, the year she married Mayes, and her master's in 1954. Thereupon, she and Mayes moved to Waco, where James opened up a dental practice while she became chair of the mathematics department at Paul Quinn College, a private historically black college founded in 1872. It was not until 1961 that Malone-Mayes decided she wanted to continue her mathematical education, applying and being rejected (on grounds of race) by Baylor University, and turning instead to the University of Texas, where she was often the only black student, and the only female student, in her courses (at least those she was permitted to attend, some professors refusing to teach black students at all).

She was awarded her PhD in 1966, becoming the fifth black woman to earn a mathematics doctorate in the US. Her subsequent research included investigations of what functions have a periodic steady state, the properties of generalised ordinary Hausdorff matrices, and the functional analysis of non-linear operators.

Alexandra Bellow (née Bagdasar) (b. 1935)

Alexandra Bagdasar was born on 30 August 1935, in Bucharest, Romania, to a psychiatrist mother and neuroscientist father. Her mother had wanted to become a civil engineer, like her father (Bellow's grandfather), but the Polytechnic Institute of Bucharest did not admit women at the time, and so she chose a medical path instead. Her parents encouraged her interest in mathematics even as their home country seemed to be falling apart all around them under the totalitarian leadership of Gheorghe Gheorghui-Dej, whose Communist regime imprisoned between 60,000 and 80,000 political dissidents, including Bellow's mother, who came under scrutiny for advocating for Western food support after the Second World War.

During this time of great personal strife, mathematics served as a balm and centre of sanity and precision in a country that rewarded ideological rectitude over truth. She attended the University of Bucharest, where she received her master's degree in 1957, a year after marrying her first husband, fellow mathematician Cassius Ionescu-Tulcea, who had been her analysis professor. Together, they moved to the United States in 1957, where he had an appointment at Yale and she worked towards her doctorate, producing her thesis on ergodic theory in 1959. She remained at Yale until 1961, later working at the University of Pennsylvania and University of Illinois at Urbana-Champaign until she found her permanent academic home at Northwestern, where she taught from 1967 to 1996.

Before we get to Bellow's work in mathematics, a few words about names: Bellow changed her name to Ionescu-Tulcea upon marrying Cassius, then changed it again, 'under duress', when she married the Nobel Prize-winning writer Saul Bellow in 1974, but refrained from changing it a third time upon marrying the mathematician Alberto Calderon in 1989, and now advises young women in science to learn from her mistakes and keep their maiden names upon marriage.

Bellow's work centred upon questions in lifting theory (where her work with her first husband established important results about how to represent linear operators in probability), ergodic theory (which investigates what happens to the particles or elements in systems or series that are allowed to run for long periods of time), and probability (with particularly interesting work in the field of martingales, which describe situations in probability whereby the expected next value of a certain amount is equal to its current value – for example, if you are rolling a dice and win \$5 if it comes up even, and lose \$5 if it comes up odd, and you currently have \$70, then your expected amount of money you'll have in your wallet after the next dice roll is \$70, because your chances of winning \$5 are exactly balanced out by your chances of losing \$5. Bellow looked at martingales in the context of a type of vector space with a way of measuring distances called a Banach Space.)

Lenore Carol Blum (née Epstein) (b. 1942)

Lenore Blum spent her youth in two very different worlds, the first nine years of her life amongst her extended Jewish family in New York, and the next seven in Caracas, Venezuela, where her father struggled to make ends meet. It was not until her mother found work as a teacher that the family was able to earn enough to send Epstein to school, where she met the young man who would eventually become her husband, Manuel Blum.

Graduating in 1958 at the age of 16, Lenore applied to MIT, where Manuel was studying, but was rejected, and settled instead for the Carnegie

Institute of Technology in Pittsburgh as an architecture student, before transferring halfway through her undergraduate career to Simmons College as a mathematics student. She received her bachelor's in 1963, and received her PhD in 1968 from MIT for her paper 'Generalized Algebraic Structures: A Model Theoretical Approach'. She then went to UC Berkeley, where Julia Robinson was teaching at the time, as a lecturer, but lacking any tenure track positions, she opted instead for a position at Mills College, a women's college in nearby Oakland founded in 1852. Starting at Mills in 1973, she wasted no time in putting together a Mathematics and Computer Science Department in 1974, the only such programme at a women's college at the time.

During the late 1980s, she turned away from administration and towards research, particularly concentrating on topics of computability, collaborating with Michael Shub on the development of the Blum-Blum-Shub pseudorandom number generator, which is based on generating the next number in a sequence by taking the square of the previous term, and looking at the remainder when that is divided by a product of large primes. This is still a popular generation method to make a number sequence that appears totally random to the naked eye. They also created a computation model called the Blum-Shub-Smale Machine. It is a clever idea for a machine that calculates over the real numbers, instead of classical Turing machines, which get distinctly uncomfortable when asked to deal with uncountable quantities like those that arise from the real numbers. In 1998, Blum, Smale, Shub, and Felix Cucker published a book encapsulating their ideas about real-number-based computation, *Complexity and Real Computation*, one year before the twenty-fifth anniversary of her founding of the Computer Science department at Mills College, and a mere six years before she was awarded the Presidential Award for Excellence in Science, Mathematics and Engineering Mentoring.

Margaret Dusa McDuff (née Waddington) (b. 1945)

Margaret Dusa Waddington was born into a rich web of intellectual accomplishment – her mother was a noted Scottish architect, her father was Conrad Hal Waddington, one of the century's great biological minds and a founder of epigenetics, her grandmother was the feminist author and scholar Amber Reeves, and her sister is the anthropologist Caroline Humphrey, who studied Soviet collective farms, and Inner Asian environmental conservation efforts in the mid- to late twentieth century.

She received her bachelor's degree from the University of Edinburgh, and then proceeded to our old friend, Girton College, where she studied functional analysis with George A. Reid, completing her doctorate in 1971. The work for which she is best known deals with the field of symplectic geometry/topology, which looks at differential manifolds (places that look like Euclidean space locally, and where you can do calculus) equipped with a measuring tool called the symplectic form. Unlike in Riemannian geometry, where you use metric tensors that take two tangent vectors at a point and output a single value that tells you information you can use to calculate the distance along a curve on that space or the angle between those vectors, in symplectic geometry, the symplectic form takes two vectors and outputs the oriented area described by them, providing new ways of looking at, and solving problems, associated with manifolds. McDuff's book, *Introduction to Symplectic Geometry*, co-written with Dietmar Salomon, is currently in its third edition.

Sun-Yung Alice Chang (b. 1948)

Chang was born in Xi'an, China, but grew up in Taiwan, where her position as the first student in her high school gave her *bao song* status, which allowed her to choose which college she wanted to attend without having to take the usual entrance exams. She chose Taiwan National University in Taipei, where she received her bachelor's degree in 1970. She then attended UC Berkeley for her graduate work, receiving her PhD in 1974. From 1980 to 1998 she was a professor at UCLA, and since then has served as the Eugene Higgins Professor of Mathematics at Princeton University. Her work has concentrated on using partial differential equations to analyse geometric structures, including an interest in fully non-linear equations and Einstein manifolds.

Cheryl Praeger (b. 1948)

Australian mathematician Cheryl Praeger is the author of an astonishing 440 mathematical papers. She was born in Toowoomba, in Queensland, and went to school at Humpybong State School in Margate. (In general, I think we'd have a better society if we had more schools with names like Humpybong and fewer with names like Vanguard Elite Excellence Academy.) Enrolling in the University of Queensland, she took courses in physics and mathematics, ultimately majoring in the latter, receiving her degree in 1969.

She wrote her first published paper while still an undergraduate, solving a problem posed to her by group theorist Bernhard Neumann (husband of Hanna Neumann, whom we met earlier), to find solutions to the functional $x(n+1) - x(n) = x(x(n))$. After graduating, she attended St Anne's College at Oxford on a Commonwealth Scholarship, where she was awarded her master's in 1972 and PhD in 1974. She then returned to Australia, where she took up a post at the University of Western Australia, where she continues to work today.

Her output over the last half century has been prodigious, and defies ready summarisation, but broadly speaking her research interests include group actions (where you take a group, and use it as a basis for creating different transformations of a set, in a way that preserves compositions), computational group theory (particularly algorithms for group computations), algebraic graph theory, and combinatorial design. She is regularly in the top 1 per cent of the world's most cited mathematicians.

Shihoko Ishii (b. 1950)

When Shihoko Ishii was a girl, her parents were worried about her 'impossible dream' of becoming a mathematician. There were no prominent women mathematicians in Japan at the time, and they were concerned that attempting to pursue a career in that field would only lead to frustration and isolation. Ishii was strongly compelled, however, to get at the mathematical roots of the structures that drive reality and, lacking inspiration of any Japanese mathematician role models, took it instead from Japan's theoretical physicist Fumiko Yanezawa, whose books encouraged young women to enter the sciences.

She received her bachelor's degree from Tokyo Women's Christian College in 1973, followed by a master's from Waseda University in 1975, following which she got married and had her child, which put serious time constraints on the furthering of her research; it was not until 1983 that she received her PhD from Tokyo Metropolitan University. Her field of research is the study of singularities of algebraic varieties by looking at the arc spaces defined by those algebraic varieties; in other words, using the set of little pieces of a curve or surface to illuminate the nature of the places where that curve or surface crosses over itself.

Shihoko Ishii received the Saruhashi Prize in 1995, eleven years after her role model, Fumiko Yanezawa, received the same award.

Annie Cuyt (b. 1956)

Annie Cuyt is a computational mathematician who earned her PhD from the University of Antwerp in 1982 for her work on Padé approximants, which are ways of approximating functions using rational functions, instead of the usual polynomials we use in Taylor Series approximations. Using rational functions often produces convergent results in places where Taylor Series cannot, and so Padé approximants are highly useful in computing. Her research also includes work on continued fractions, which are essentially formed by stuffing fractions inside of fractions iteratively, resulting in sprawling infinitely fractioned quantities that cause a sensation of churning distress in anybody who toiled through the fraction-obsessed United States elementary school mathematical curriculum of the last fifty years.

Irene Fonseca (b. 1956)

Portuguese mathematician Irene Fonesca is the author of multiple papers on the application of the calculus of variations to the analysis of situations and objects that arise in materials science, including phase transitions, thin structures, magnetic materials, shape memory alloys, and composites. She attended high school in Lisbon, before earning her Licenciatura in Mathematics from the University of Lisbon in 1980, then doing her graduate work on a Fulbright Fellowship at the University of Minnesota, where she received her PhD in 1985. She currently directs the Center for Non-linear Analysis at Carnegie-Mellon.

Małgorzata Klimek (b. 1957)

Born in the southern Polish city of Bielsko-Biała in 1957, Klimek earned her master's at the Wrocław University of Science and Technology in 1981, and her PhD at the theoretical physics department of the University of Wrocław in 1993. She is currently a professor at the Częstochowa University of Technology, where her research interests include fractional mechanics and fractional differential equations, and the investigation of what conservation laws hold for different noncommutative spaces.

172

Éva Tardos (b. 1957)

An expert in algorithm theory and design, the Hungarian-born Éva Tardos is currently the Associate Dean of the College of Computing and Information Science at Cornell University after a half decade as the Chair of the Computer Science department there. Her speciality is the design of algorithms that produce close-to-optimal results in game and auction theory, and the development of approximation algorithms for phenomena like network flow (directed graphs which have limits on how much flow can pass through an edge and which therefore approximate real-life situations where quantities must flow through different possible channels subject to upper limits on flow rates), and clustering tasks (ways of automatically arranging groups of objects so that similar objects end up being near each other, or certain quantities are minimised, as in the Facility Location Problem). In 2005 she and Jon Kleinberg published the book *Algorithm Design.*

Zhilan Julie Feng (b. 1959)

Zhilan Feng is a specialist in a branch of mathematics that has grown very relevant in recent years: the mathematical modelling of issues related to public health and epidemic events. She received her bachelor's degree in 1982 and master's degree in 1985 from Jilin University, before receiving her PhD from Arizona State University in 1994. She is currently a professor of mathematics at Purdue University, where her work in the twenty-first century has dealt with modelling disease evolution, vaccine effectiveness over time, different strategies for addressing epidemics, and the dynamics of virulent outbreaks within a host. In 2019, she published *Mathematical Models in Epidemiology.*

Farideh Firoozbakht (1962–2019)

Farideh Firoozbakht is known for her 1982 conjecture that the nth root of the nth prime keeps getting smaller as n increases, i.e. that the first root of 2, the first prime, is greater than the second root of 3, the second prime, which is greater than the third root of 5, the third prime, which is greater than the fourth root of 7, the fourth prime, and so on. Known as Firoozbakht's Conjecture, it has yet to be proven, but has been checked and found true up

18 quadrillion, and if true, it implies some interesting things, such that there are always two primes in between any two consecutive square numbers and four primes between any two consecutive cubes.

Guofang Wei (b. 1965)

UC Santa Barbara professor Guofang Wei was born in the People's Republic of China, where she received her master's degree from Zhejiang University, before moving to the United States for her graduate work at SUNY at Stony Brook in New York. Her research centres around global Riemannian geometry, including much work on the topology of manifolds with different Ricci curvatures. She also worked with Christina Sormani in 2004 to develop the notion of a 'covering spectrum' of a Riemannian manifold, which gives you a way of measuring the size of any holes in the space.

Donatella Danielli (b. 1966)

Danielli received her bachelor's degree from the University of Bologna (1989) and her doctorate from Purdue (1999), where she served as a faculty member until transferring to her current posting at Arizona State University. Her early research included various studies of non-linear sub-elliptic equations, while recently she has concentrated on investigations into free boundary problems arising at the interface of different phases of matter, in particular involving flame propagation. She won the National Science Foundation CAREER award in 2003, and was made a fellow of the American Mathematical Society in 2017.

Clara Grima (b. 1971)

Computational geometer and mathematics communicator Clara Grima is known equally well in professional communities for her research on scutoids (which are geometric solids that resemble prisms, but possess at least one Y-shaped edge which allow them to completely pack in the surface between two parallel planes), and in popular communities for her books on the connections between mathematics and personal health, and the general role that mathematical literacy can serve in enriching one's everyday understanding of the world.

Anna Catherine Gilbert (b. 1972)

Gilbert is a professor at Yale University and the 2013 Ralph E. Kleinman Prize winner whose work unites the kingdoms of mathematics, electrical engineering and computer science. Her particular fields of interest include the application of randomised algorithms for the efficient handling of large data sets and processing of signals and images, and the use of sparse analysis in signal processing and network traffic analysis.

Marta Lewicka (b. 1972)

Author of nearly seventy papers spread over half a dozen different realms of mathematics, Lewicka, who earned her bachelor's and master's degrees from the University of Gdansk, and her PhD from the International School for Advanced Studies in Trieste, is living proof that the future of mathematics need not belong to the hyper-specialist. She has published work concerning the calculus of variations, non-linear elasticity, the analysis of prestrained materials, game theory, fluid dynamics, and continuum mechanics.

Amandine Aftalion (b. 1973)

Since 2008, Aftalion has served as a director of research at the Centre Nationale de la Recherche Scientifique (CNRS) in France, where she has performed her investigations in applied mathematics since 1999. Her 1997 dissertation centred on elliptic partial differential equations and their application to the phenomena of superconductivity, and her subsequent work has investigated the Ginzburg–Landau equations of superconductivity, and the problems associated with two component Bose–Einstein condensates (collections of bosons at a temperature near absolute zero, which can manifest macroscopic quantum phenomena). In 2006, she wrote a book on Bose–Einstein condensates, studying their vortex and superfluid behaviours. In a seemingly completely different vein, she has applied differential equations to the study of the forces and anaerobic/aerobic activity arising from humans undergoing long-distance runs.

Ilka Agricola (b. 1973)

Ilka Agricola is a differential geometry specialist whose work focuses on variously exotic manifolds, the spinors and torsions associated therewith, and their connections to ideas in physics such as string theory, and supergravity. (Manifolds are topological spaces that are locally similar to our old friend Euclidean space and therefore allow us to use some of our favourite mathematical tools there, while spinors are elements of complex vector spaces that respond to overall rotations differently depending on the order of the steps taken to arrive at those rotations.) She received the Humboldt Prize in 2000, the Medal of Honor of Prague's Charles University in 2003, and was elected president of the German Mathematical Society in 2021.

Daniela Kühn (b. 1973)

Daniela Kühn is a German mathematician who received her PhD from the University of Hamburg in 2001, and a recipient of the Richard Rado Prize (2002), the European Prize in Combinatorics (2003), and Whitehead Prize (2014). Her work centres on extremal combinatorics and graph theory, with many of her published works centring around Hamilton cycles of directed graphs, which are paths through a collection of vertices that hit every vertex once, and end where they began, and hypergraphs, wherein edges are allowed to connect more than two vertices.

Erika Tatiana Camacho (b. 1974)

Born in Guadalajara, Mexico, Camacho grew up in Los Angeles, where she attended Garfield High School from 1990 to 1993, learning mathematics in classes taught by Jaime Escalante, who movie buffs might remember as the inspiration for the film *Stand and Deliver*. She graduated from Wellesley College in 1997, and earned her PhD in applied mathematics from Cornell in 2003 for her work on the mathematical modelling of retinal decay. Since then, her work has shed light on the mathematical mechanisms underlying various biological processes, from fungal infection in immuno-compromised individuals, to gene regulatory networks, to aerobic glycolysis, all while working to increase Latina representation in STEM.

Nina Snaith (b. 1974)

Nina Snaith is a researcher in Random Matrix Theory, with a concentration in investigating the statistics of the eigenvalues and characteristic polynomials produced by random matrices. In 1998, she and Jonathan Keating used Random Matrix Theory to conjecture a value for the constant coefficient that approximates the moments of the Riemann Zeta Function, which was subsequently found, using entirely different methods, by Diaconu, Goldfeld and Hoffstein. She has since gone on to apply RMT to illuminating how events along chromosomes are distributed.

Kathrin Bringmann (b. 1977)

Bringmann is one of the most decorated of young professors in the mathematical world today, having won the SASTRA Ramanujan Prize in 2009, the Alfried Krupp-Forderpreis for Young Professors, also in 2009, and the 2018 Prose Award for best scholarly book in mathematics from the Association of American Publishers. So, what is all the fuss about? Bringmann's field of study is Mock Theta Functions, which were first described by the great Srinivasa Ramanujan in 1920. In the century since that description, much work has been done attempting to exactly connect Theta Functions and Mock Theta Functions, to little avail until the twenty-first century when mathematicians like Sander Zwegers, Ken Ono and Bringmann found ways to drag the mock theta functions into the realm of modular forms.

Miranda Cheng (b. 1979)

Miranda Cheng took the less-trodden path to a life of mathematical study by way of punk rock. Born in Taiwan, she found herself consistently uninterested in, and unchallenged by, her schoolwork, attending school as little as possible and ultimately dropping out and running away from home at the age of 16 to work in a record store and join a punk band. The chaos of a musician's life, however, soon called for something calm and profound to counterbalance it, and she found herself attending university, irregularly at first until she found a class in quantum mechanics that profoundly affected her. In 2001, she moved to the Netherlands to carry on her studies at Utrecht

University and the University of Amsterdam. In 2012, she, John Duncan and Jeffrey Harvey formulated the Umbral Moonshine Conjecture, which proposed twenty-three new moonshines, which are mathematical structures that link symmetry groups and mock modular forms, and which should help elucidate the properties of the K3 model of ten-dimensional string theory.

Sophie Morel (b. 1979)

Sophie Morel's mathematical life began when her mother, a school librarian, brought home one day a copy of the high-school oriented mathematical magazine *Tangente*. It taught maths that was not part of Morel's normal high school curriculum – weird maths, fun maths, maths to be wrestled with, and Morel immediately took out a subscription. That experience in maths beyond the classroom led to others, including a mathematical summer camp she attended, and hours of leisure time spent reading upper division mathematical texts for the sheer joy of learning things that were not based on the solve for x grind of the official curriculum.

Today, Morel is a number theorist who won a 2012 European Mathematical Society Prize, and in 2009 became the first tenured woman professor of mathematics at Harvard University. Her work centres around the Langlands programme, a project formed by Robert Langlands in the late 1960s to connect number theory to representation theory, and she has written a number of papers on Shimura varieties.

Corinna Ulcigrai (b. 1980)

Corinna Ulcigrai, an Italian mathematician who received her PhD in 2007 at Princeton, came to prominence in 2013 when she and Krzysztof Fraczek proved that most trajectories in an Ehrenfest space are not ergodic. The Ehrenfest space (often called the Ehrenfest Model, but which I'm renaming here to avoid confusion with the dog-flea model) is due to our friends Tatyana and Paul, discussed previously, and consists of an infinite plane containing an infinite amount of rectangular obstacles regularly placed, which the Ehrenfests used to analyse the statistical behaviour of gases. Prior to Ulcigrai and Fraczek's proof, mathematicians generally believed that a particle set loose in a random direction would follow an ergodic path, i.e. it would, at some time or another, visit every location in the plane. Ulcigrai and Fraczek found that the most trajectories, in fact, are not ergodic, a result that won Ulcigrai the prestigious Whitehead Prize in 2013.

Sara Zahedi (b. 1981)

Born in Tehran, Sara Zahedi was sent to Sweden at the age of 10 after her father was killed in the aftermath of the 1979 Iranian Revolution. She earned a master's degree from the Royal Institute of Technology in 2006, and a PhD in 2011. In 2016, she received a European Mathematical Society Prize for her numerical analysis of systems with dynamically changing geometry, such as the fluid dynamics problems one comes across when trying to deal with the changing boundary conditions of two immiscible fluids that come in contact.

Irina Shevtsova (b. 1983)

Author of over seventy papers, Irina Shevtsova is currently a professor at Moscow State University, where she began working in 2006 and earned her PhD in 2013. One of her main areas of research is methods for approximating the rate of convergence of sums of independent random variables. The Central Limit Theorem of probability states that independent random variables, when summed, have a distribution that tends towards the classic normal distribution. Shevtosva's work investigates how to improve estimates for convergence towards the normal distribution.

Rachel Riley (b. 1986)

When you sit down at a table, and start doing mathematics, no matter what the level, you instinctively *know* that you are doing something Pretty Damn Cool. Seeing a problem, breaking it down into its component bits, finding a method that will get at the solution, and then coming up with a number that matches reality – that is just fundamentally clever. The problem is, a lot of kids cannot quite visualise that the cool things they are doing can be done by somebody who is, themselves, cool. We believe instead that, to enter into the domain of mathematics and enjoy its surprises and delights, we have to give up any role in the world, and devolve into some manner of troglodyte.

It is a frightening, lonely, prospect to many, which has turned more than one person away from the gates of deeper mathematical enjoyment.

Enter Rachel Riley. Since she surpassed 1,000 other applicants in 2009 for the role of number wrangler on the popular British television programme *Countdown*, and parlayed that into a similar role on its edgier cousin *8 out of 10 Cats Does Countdown* in 2012, she has stood as an example and inspiration of how the coolest, most popular person in the room can also be the 'mathiest' person in the room. And not just privately, secretly 'mathy', but openly, enthusiastically 'mathy'. She has broken long-standing stereotypes and reassured a generation that the gates of mathematics are the beginning of a journey, and not the end of their social lives, and that has been a service to mathematics, its inclusivity, its demystification and popularisation, every bit as weighty as the other accomplishments listed in this volume.

In 2021, Riley published her first book, *At Sixes and Sevens: How to Understand Numbers and Make Maths Easy*.

So, What Precisely
Does that ... Mean? A Glossary of
Mathematical Terms

The wonderful, terrible thing about mathematics is how precisely every single word and symbol is defined, such that something, once established, remains established, impervious (mostly) to the passing of time. The problem is, that to define things so precisely requires a slew of new, technical words that look intimidating, but often stand for concepts that can be grasped, in their essence, without too much trouble. Some of these words helpfully point the way to what they might mean ('continuity' 'connectedness' 'combination') while others distinctly do not ('Cauchy sequence' 'Fourier Series' 'manifold'). This guide of terms that come up relatively often in this book is meant to give you the flavour of each concept without leading you into the treacherous weeds of mathematical notation and dense terminology.

Algebra

Mathematicians can mean a few different things by the word 'algebra'. The definition most of us regular humans mean when we say algebra is what we learned in school – a set of rules for working with equations that contain numbers and variables. But you might have wondered at some point, 'Is there a way to apply these same ideas to objects that don't look like the numbers, x's, and y's that I'm used to? Can I do Algebra-Type Stuff to other types of things?' Well, if you have, congratulations, you're on your way to Abstract Algebra, which studies alternate types of mathematical operations that 'work' on alternate types of mathematical objects in the same way that our elementary school algebra 'worked' on real numbers and unknown variables. For example, we could create a new meaning for 'equals' that says two numbers 'equal' each other if they have the same remainder when you divide them by, say, 7 (so, 12 and 47 would be 'equal' to each other under this definition, because both have a remainder of 5 when you divide them by 7). An abstract algebraist would then go on to look at what new mathematical structures result when we use this new definition of 'equals' and, perhaps even more interestingly, which structures and operations stay pretty much the same.

There is yet a third thing that can be meant when you see the word 'algebra' and that is not a way of thinking about numbers and the operations you can perform on that, but an actual mathematical object. That object consists of a set of things, a rule for how you 'multiply' objects in that set together, and a separate set of numbers that you can use to scale the objects in that set. So, for instance, in everyday life our 'algebra' has a set of things, the real numbers, any two of which elements we can multiply together using our standard multiplication, and any one of which we can change the magnitude of by multiplying by anything from the world of, once again, real numbers. But just imagine what it might be like if we tinkered with any one of those parts, and you're starting to conceive of a world with new algebras with startling new behaviours, waiting to be explored.

Analytic

We like analytic functions. Essentially, an analytic function is just a function that you can take as many derivatives of as you would ever want to, i.e. one that has behaviour that is nice and smooth no matter how deeply you look into it. Analytic functions can be rewritten as power series (i.e. as a sum of a bunch of different powers of x), which is great, because working with power series is very pleasant. So, when you see somebody mentioning that a function is 'analytic' you just need to breathe a sigh of relief as you realise that there will be no temperamental jags, corners, or vertical asymptotes to play havoc with your analysis. Of course, a LITTLE havoc now and then is a good thing too ….

Banach Space

Technically, a Banach space is defined as a 'complete normed vector space'. The 'normed' part of that just means that a Banach space is a place that comes equipped with a way to measure the 'size' of elements in the set (which is what the 'norm' does), which also implies a way of measuring the 'distance' between elements in that place. The 'complete' part is a bit trickier, but basically just says that if you have a sequence of elements in that space that are getting generally closer and closer to each other (called a Cauchy sequence), that eventually they will converge to a location that is itself within that space. This is what the word 'completeness' is trying to get at, that the space comes complete with the destination points for all the Cauchy sequences in it.

Calculus

Co-invented by Isaac Newton and Wilhelm Gottfried Leibniz in the seventeenth century, calculus is one of the most powerful set of methods for analysing the world around us yet created by the human mind. Suppose you have a graph that shows you the position of an object at different points in time. Calculus lets you ascertain, at every moment in time, how fast the position of that object is changing, and thereby gives you a way to know the velocity of that object at every moment in time, and THEN, when you build a graph out of that information, you can determine how fast the VELOCITY is changing at every moment in time, which gives you the acceleration of that object at every moment, which then opens up the ability to know the net force on the object at every moment in time. That is an absolute treasure trove of information that calculus lets you mine out of one simple position graph, but it doesn't stop there!

While differential calculus lets you analyse the rates of change of equations and graphs, another part, called integral calculus, provides scientists with powerful tools for dissecting the phenomena of the natural world. For example, suppose you want to know the total gravitational impact of a star at a point a given distance away from that object. Well, according to Newton's law, gravity is an inverse square relationship, so two points on that star that are a different distance away from the given point will have different gravitational impacts there. What integral calculus lets you do is slice up the star into a bunch of infinitesimally thin regions that are all the same distance from the point, evaluate the total gravitational impact of each of those slices, and then add up all infinity of those infinitely thin slices to determine the star's total force of gravity at that point. That sounds like dark wizardry, but Newton and Leibniz found a way to make it work by harnessing the power of limits, and the result is the world we see around us.

Closed/Open Set

This is a key concept in a lot of different branches of mathematics which, seemingly just to drive us all crazy, all have slightly different ways of defining just what it means. Let's start our investigation into this idea by looking at our old friend, the number line. An 'open set' on the number line is just an interval or collection of intervals wherein each point has some 'room' around it that contains only points also in the set. For example, the open interval (0,1) is an open set, because if you give me any point in that set, say .95, I can give you a radius around that point, say .01, that will only contain other numbers in that set. The closed interval [0,1], however, is not an open set because it contains the points 0 and 1, and if you choose the number 1 from that set, I am totally unable to give you any radius around 1 that is guaranteed to only contain numbers from that set (there is no 'room' to the right of 1). This is what, on its most basic level, defines open and closed sets in the everyday world that we know – if every point has a radius of in-set neighbours around it, it is open. If it doesn't, it is not open (maddeningly, this doesn't necessarily mean that it is closed, it just means that it is Not Open. Closed sets are defined as sets whose complements in the space are open sets, i.e. I am closed if the set containing everything that isn't me is open – therefore [0,1] is closed because (-infinity, 0) U (1,infinity) is open).

Topology relies heavily on the idea of generalised open sets, because it studies how the properties of a space change as we change how we define the open sets in that space. Part of the fun of topology is looking at ways to define an 'open set' that might look very different from our well-behaved intervals on the number line, but we'll learn more about that below when we talk about topology.

Combinatorics

If you have ever asked 'How Many Ways Are There To Do This?' then you have dipped your toe into combinatorics. How many ways are there to make a twelve-song playlist from a set of thirty songs? How many ways are there to draw five cards from a standard fifty-two card deck? How many different ways are there to get from point A to point B, if I have to make a series of decisions along the way of one of several minor roads to take? One of the big distinctions you come across in beginning combinatorics is the notion of Order. Whether or not order matters in a particular situation has a profound impact on how many ways there are to do something. If I'm just trying to pick three people from a group of twelve to throw in a car for a road trip, it is all the same to me whether I pick the group {Tatyana, Charlotte, Sofia} or {Charlotte, Tatyana, Sofia} – the order that they get into the car isn't important to me, and so I consider any group of the same people to be equivalent. When order doesn't matter, we talk about each possible group as a Combination. When order does matter (when, say, I'm trying to award 1st, 2nd, and 3rd place in a sports competition, and therefore {England, Algeria, Mexico} is considered to be a VERY different thing than {Algeria, Mexico, England}) then I talk about each group as a Permutation. Generally speaking, if I'm trying to form groups of size m by taking elements from a larger group of size n, there will always be as many or more Permutations than Combinations, because there will be fewer groupings getting lumped together as 'the same group'.

Connected / Path-Connected / Simply Connected

In Topology, one of the things that we care about when looking at a space is its connectedness, and what happens to that when we start messing with the space in variously perverse ways. Put crudely, a space is connected if I cannot find a way to contain it within two open sets that don't intersect with each other. Visually, in the everyday world of Euclidean space using normal ideas of what open sets are, this boils down to a space being connected if there are no gaps that cleave it into two separate regions.

There are more restricted forms of connectedness that you'll see when people are listing the properties of the space they want to talk about. My favourite is Path-Connectedness, which basically just means that, if you take any two points in the space, there is a way to get from one to the other whereby every step of your journey lies within the space. Being path-connected means that a space is necessarily connected, but being connected doesn't force it to be path-connected, and there are some fun examples of spaces that are connected but not path-connected that mathematicians have cooked up that you should check out.

Remember when I said that path-connectedness was my favourite type of connectedness? I lied. It is actually simple connectedness. A region that is simple (or simply) connected is, visually, one that has no gaping holes in it. More formally, it is a space where, if you take two points in the space, and two different paths between those two points in the space, there will be a way to 'drag' the one path over to the other path so that, at each point during the dragging process, the path stays completely within the region.

Continuity

For students whose love of mathematics survives their introduction to logarithms, the next great rite of passage is often their introduction to the epsilon-delta definition of continuity. For some, it is an elegant means to formally define something that had been hitherto only loosely spoken of, and is the gateway to a whole career of finding ways to encapsulate everyday ideas in terms so that they can be applied in more abstract worlds. For others, it is the end of the road, the moment that they decide, 'Well, I guess I'm not going to turn out to be a mathematician then.'

Firstly, I'd say don't ever let not 'getting' a mathematics concept prevent you from pursuing mathematics – there are acres and acres to explore in the mathematical universe, and part of the fun is finding out the part of that universe that clicks best with how your brain sees the world. Secondly, I'd say that, beneath all the frightening Greek letters, the epsilon-delta definition of function continuity is a pretty simple thing. Basically, when we say that a function is 'continuous' what we are saying is that, if we apply that function to a group of points that start out close together, they will end up close together. Continuous functions are our pals, because they ensure that neighbourhoods don't get broken up. For instance, the function $f(x) = x^2$ is a continuous function, because if we take a neighbourhood of x-values, say numbers close to $x=2$, and shove them into the function, they'll come out as a nice neighbourhood of numbers close to 4. The greatest integer function, however, which takes a number, and reduces it to the biggest integer that is less than it (so, for example, 4.32 gets output as 4, and -2.373 as -3), is not a continuous function, because it breaks up neighbourhoods. For numbers around $x=2$, for example, those that are less than 2 will end up getting output as '1' whereas those above $x=2$ will end up getting output as '2'.

Functions take us on a journey from one space to another space. Continuous functions make sure that we don't get broken up along the way.

Convergence

Just like we, as humans, are obsessed with our destiny, where we shall end up at the end of our life's path, so too do mathematical functions and sequences constantly wonder what will become of them. Will they ultimately settle down, getting closer and closer to a particular stable value, or will they keep oscillating back and forth between two values or, worse, head off into the unruly regions of infinity, ever growing without limit or, one might say, decorum? If a sequence or function gets ever closer to a particular value, we say it converges to that value. If it does not converge, either because it oscillates or because it capers off to infinity with a sort of Mad Max like crazed nihilism, then we say it is divergent. Generally speaking, convergent sequences are the better-behaved ones, but divergent sequences are more fun.

Derivative

Often times, when looking at a graph, we are interested in how steep it is at different points. To find out the steepness of a graph at a point, we take the derivative of the graph's equation, and plug that point into it. We can find the value of the derivative at a point on the graph as long as two conditions are met: (1) the graph is continuous there (see 'continuity' above, and if that is too alarming, just think of a continuous graph as one you don't have to lift up your pencil to draw, and that'll do for now), and (2) the graph doesn't have pointy bits on it – this is a very technical way of saying that the slope of the function doesn't suddenly and instantly lurch from one value to a totally different value. Derivatives are very important in physics, because they give us information about how changing one variable by a little bit causes changes in other, related variables.

Differential Equation

Most equations that you experience in secondary school involve just numbers and variables: $y = 2x + 3$, $x^2 + y^2 = 16$, $\sin(x) = y^2$. These might be difficult to visualise and analyse, but they are pleasant to work with in the sense that they only involve how the *value* of one quantity impacts the *value* of another quantity. So, when we 'solve' one of these equations, what we are often doing is just figuring out, for example, what values of y we get when we plug in certain values of x. Simple. Peaceful. Moral.

Differential equations throw all of that out of the door, and ask something both much more challenging and much more entertaining from us. Instead of limiting themselves to putting numbers and variables into the equation, they allow themselves also to put into the equation terms which represent different derivatives of variables. And then, instead of the solution consisting of just numbers, what a differential equation is asking us for is to create *a new equation* that meets all of the criteria put forth in the differential equation. So, for example, if I see $y'' + y = 0$, what I am being asked to do is find an equation such that, when I take the second derivative of it, and add it to itself, the end result is 0. Maths and physics students will generally spend a year of their college careers learning different techniques for finding the equation that satisfies a given differential equation. At the end of that year, they emerge from their study holes looking a bit like Batman, with a utility pouch for every possible contingency, wielding techniques to combat functions containing derivatives of any degree, and even containing those most dread of foes, *partial* derivatives. Physics is full of differential equations, because it is full of questions about how motion and acceleration and position all relate to each other under different constraints, so if you're thinking about a future in physics, start training yourself to become a DiffEq ninja now.

Euclidean Space

The nice space. This is the space you worked with throughout primary and secondary school, with its nice x, y and z axes, usually equipped with the good old Pythagorean Theorem for determining the distance between points and length of any vector in the space, as well as the dot product for figuring out the angle between two vectors in the space. This space has the same density everywhere, which is also pleasant. Often, in maths, a good way of coming to grips with a new idea is to think about how it works in well-behaved, easily visualised Euclidean space, and then start generalising outward from there to weirder spaces.

Ergodic

An ergodic system is, in its most basic essence, just a space wherein, if a particle is free to roam, and given enough time, it will eventually visit every point within the space. A good first pass at visualising ergodic systems is to think of idealised gas particles introduced into a jar – those particles will expand to fill the space, and will over time explore every part of it. An example of a non-ergodic system would be a group of low-energy balls in a plane that has several large pits in it. Any low-energy ball that starts out in a pit will lack the energy to get themselves out of it, and so will be stuck exploring the pit they happen to be in for the rest of time, without getting to roam to the other parts of the space.

Field

When you think about working with numbers, you're fundamentally thinking about two things: (1) the set of numbers you're allowing yourself to work with, and (2) the types of ways that you're allowing yourself to combine those numbers. In mathematics, we have different words to describe different mixes of number sets and combination rules. The one that most familiarly approximates numbers as we work with them on a day-to-day basis is the field.

A field is just a set of elements, equipped with a way to add those elements together, and a way to multiply them together. That set must contain an 'additive identity', i.e. a number that works like '0' does in our normal arithmetic, whereby, for any element a in the set, $a + `0' = a$. It must contain a multiplicative identity that works like our 1 does, such that $a \times `1' = a$. Further, the addition and multiplication rules we create must be associative and commutative (so switching their order, or how we group them doesn't impact the end result of an addition or multiplication), and the set must be 'closed' under addition and multiplication, which just means that if you add or multiply any two elements of the set, the result is a number also in that set.

The real numbers, Rational numbers and complex numbers, equipped with everyday addition and multiplication, are all classic examples of Fields, but mathematicians delight in creating much more wacky examples that stay within these rules but produce engagingly unfamiliar number worlds.

Fourier Series

Think of the last science fiction movie you saw where the villain communicates to his subjects through an audio device that shows the wave pattern of his voice while he is talking. That wave changes its shape constantly as he is talking, and if you were to look at the shape of that wave over time you'd get a dizzying array of peaks and valleys of seemingly random widths and heights. One of the challenges of digital telecommunication was finding ways to take that crazy wavy pattern, digitise the information contained in it, and then reproduce enough of that information at another location so that the sound coming through the telephone on the other end sounded enough like the speaker's voice for their phone partner to convince themselves that what they were hearing was actually their friend, and not the robot cleverly using maths to mimic their friend that they were actually hearing.

What allows us to analyse curves and phenomena like this are Fourier Series. If you picture an ordinary sine or cosine curve, what you're seeing is already an object that has a good amount of curviness to it. What Fourier Series do is to take a whole bunch of sine and cosine curves, stretch them out or shrink them down, and then add them together to approximate a given curve or phenomenon. The more sines and cosines you throw in, the more accurately you can reproduce the target curve, but the more information you'll have to send, so the question is always striking the right balance between fidelity and the cost of information transmission.

Function

When you first work with equations in elementary school, you usually think of it as a rule for changing inputs to outputs. The equation $y = 3x + 2$ is a relation that lets you change the inputs, x, into outputs, y, by tripling them and adding two. A 'relation' is a way of generating outputs from inputs that has no particular restrictions on what the outputs are allowed to be. A 'function' is a relation with one important restriction, which is that, for each input, there is only allowed to be one output. So, if I have a function that tells me how much money I'm making by selling x Thor action figures, and I input 300 for x, I know that that function is going to give me a single number as an output, whereas for a relation, I might get multiple different outputs for plugging 300 into it, which makes business planning, to put it mildly, difficult.

Group

Remember Fields from above? How that was a set of numbers with an 'addition' and 'multiplication' operation that followed certain rules, with a couple of identity elements thrown in, and the general idea that the set was closed under addition and multiplication? Well, good news, groups are like those, but simpler!

A 'group' is just a set of elements that comes equipped with an operation, usually denoted with a $*$, that obeys three restrictions: (1) The operation is associative: $(a*b)*c = a*(b*c)$. (2) There is an identity element in the set, e, such that $a*e = a$. (3) Every element in the set, a, has an inverse in the set, b, such that $a*b = e$.

The integers, using the operation of addition, would therefore be considered a group, with the identity element being 0, and the inverse element of a being $-a$, but of course we can take this abstract idea of a group and import it into lands very different than the set of numbers, including the world of polynomials, or matrices, or geometric transformations.

> **Abelian Group:** This is just a group with the added feature that the group's operation is also commutative, i.e. $a*b = b*a$.

> **Finite Group:** This is a group whose set only has a finite number of elements in it. For example, the group $\{i, -1, -i, 1\}$ is a finite group under multiplication, with 1 as the operational identity. This particular group is further called a 'cyclic' group generated by i because you can get any element in the set by raising i to different powers, and as you cycle through different powers of i, you just keep regularly repeating the elements in the set.

> **Lie Group:** Remember when I said that groups can be made out of some interesting things? Well, a Lie Group is a group that also happens to be a differentiable manifold, i.e. where the set of elements that make it up is a space that, when you look closely at it, resembles peaceful Euclidean space, and that has a multiplication operation on it for combining two points in the space that produces 'smooth' results (which is the 'differentiable' part). The mathematician Sophus Lie used these to study symmetries of differential equations in the nineteenth century.

Homomorphism

Let's talk about mappings. The good news is, you did these all the time back in school. Whenever you plugged numbers into $y=x^2$ you were performing a mapping, which took an element from the Real numbers, which you called *x*, and transformed into an element also from the real numbers, which you called y, using the mapping $x \rightarrow x^2$. You could also create a mapping from, say, the number line to the *x*–*y* plane that takes a point *x* and moves it to (*x*,*x*+2), or from the set of 3rd degree polynomials to a point on the real number line, like $a_0 + a_1 x + a_2 x^2 + a_3 x^3 \rightarrow a_0 + a_1 + a_2 + a_3$. You *could* make all sorts of crazy mappings from places to other, related places, but what mathematicians are often interested in are mappings that preserve particular aspects of the original space, and port them over in some way to the target space. A homomorphism, then, is a mapping that preserves the structure of the two zones' operations. So, for example, if you have a mapping $f(x)$ from space *A* that has a 'multiply' operation denoted by *, and are porting those elements into space B that has a 'multiply' operation denoted by \$, then if *a* and *b* are in *A*, $f(a*b) = f(a) \$ f(b)$. What this says, then is that if you perform the operation * to two elements in *A*, and then map them over to *B*, you'll get the same result as if you mapped *a* & *b* over into *B*, and then performed \$ on the results.

There are lots of other restrictions you can place on a mapping, each of which (of course) has its own name. Isomorphisms are mappings that you can reverse with an inverse mapping. Automorphisms are mappings that take elements from a space and map them to other elements in that space, while preserving structure. And so on. All of the morphisms, taken as a family, represent a useful set of tools that let you move between different seeming, but structurally similar, mathscapes. What they let you do is take problems that are hard to solve in one mathematical universe, port them over to a universe where they might be easier to solve (because you have a greater grasp of problem-solving techniques in that universe, for example), and then port the solutions back to the original zone (if your mapping is an isomorphism). This is a tremendous tool in the mathematician's toolbox, and it is all thanks to an old friend most people don't know they have.

Integer

These are the friendly numbers. -2, -1, 0, 1, 2, and so forth. No fractions. No crazy non-repeating decimals, just good wholesome counting numbers and their opposites. Plumbing the depths of the properties of the integers makes up a good part of the work of Number Theory, which concerns itself with all sorts of interesting questions about primes, convergence, powers and divisibility, a treasure trove of which you can find over at the Numberphile YouTube channel, where I spend more time than I probably should.

Manifold

When you look up images of manifolds from those hot topology websites we all frequent, what you usually see are intimidating, twisted surfaces that defy you to make sense of what is happening to them. And such they might be, were it not for a very important property of manifolds, which is that they are spaces that, if you focus your attention to their individual points, have neighbourhoods that behave, structurally, the same way our hero Euclidean space does. They are like chaotic countries that are populated entirely by peaceful, law-abiding towns. The simplest example you generally come across of a manifold is a circle – Yes, it is curvy-curvy everywhere, but if you zoom in to a point on the circle, and consider all of the points within a certain distance of that point, you'll find that you can map those points over to the regular Euclidean number line, answer questions you had about the circle there, and then move the answer back to the original circle through the inverse of your mapping. There are manifolds that have even more peaceful towns, which allow you to do calculus operations in their neighbourhoods, or compute distances in relatively normal ways. The combination of local regularity with global strangeness make manifolds interesting and popular objects to work with.

Mathematics

The Most Beautiful Thing That There Is.

Matrix

At their most basic, matrices are just rectangular boxes with numbers shoved into them, that look kind of like this:

$$\begin{pmatrix} 2 & 3 & 5 \\ 3 & 7 & 10 \end{pmatrix}$$

They are used all over the place in maths, in a variety of different ways. The above matrix could, for example, be used to represent the system of equations:

$$2x + 3y = 5$$
$$3x + 7y = 10$$

By performing different allowed manipulations of the matrix, you can work it into a solution (x, y) of the system of equations, without the need for the algebraic manipulation you generally employ when solving a system of equations. Another use of matrices, however, is as mappings. You can encode into a matrix information for how to take a point from a space and manipulate it to become a member of a target space with certain properties. Then, to map a point x from space A into space B, all you need to do is multiply the coordinates of x by the mapping matrix you custom made for the job. In geometry, you often come across matrices that perform different transformations, so for example if you want to take the point (2,3) and rotate it by 90 degrees clockwise around the origin of the x–y coordinate system, all you need to do is grab your standard 'Rotates stuff by 90 degrees matrix' and multiply your point by that, and the output will be a new point, rotated exactly 90 degrees from your original. They are wonderful tools that take things that we often do by laborious hand manipulations, and reduce those operations to banks of numbers that we can then feed into our electronic manservants, and have them just handle the number crunching to tell us what the end result is.

Linear Algebra consists largely in the study of matrices, their properties and uses, and how different mechanical aspects of a matrix correspond to different features of the reality they are constructed to represent. If you're pretty good at keeping indexed subscripts straight, it is a class well worth the taking!

Norm or Normed

When presented with a set of objects, one thing that you often want in mathematics is a way of determining their relative size, which allows us to say definitely that object x is 'bigger' in magnitude than object y. Norms are what let us do this. In Euclidean spaces, we generally talk about the magnitude of an element with reference to how far away that element is from the space's 'origin' point. So, both -5 and 5 have a magnitude of 5, because they are both 5 away from the origin. (3,4) and (4,-3) both have magnitude of 5, because if you apply the standard norm of the x–y coordinate plane (i.e. the Pythagorean Theorem), you get that both of those points are 5 away from the origin of (0,0). For most of us, that Euclidean norm, based on the Pythagorean Theorem, is the only one we'll ever use, but of course, maths being maths, there exist other ways of measuring the distance of a point from the origin, like the famous 'Taxicab' norm, which responds to the fact that the distance you have to go to get from the origin to a point in space is often longer than the straight line distance, due to restrictions on where and how you can travel. Under the Taxicab norm, the magnitude of (3,4) would be 7, because it only allows you to travel in straight vertical and horizontal segments, like a taxicab driver driving on a grid of streets.

Once a space has a norm, that usually brings along with it an inherited 'metric', i.e. a way of measuring the distance between two points in that space, which is the magnitude of the difference between the two points.

Rational Numbers

Rational numbers are numbers created when you divide an integer by another integer, like 5/7, or -31/342. One of the things maths folks really like about the Rational numbers is that they are countable, which the Real numbers are not. What this means is that you can create an algorithm that can assign to every Rational number an integer, so you can count them the way you do sheep, or matchsticks, or days until your manuscript about the history of women in mathematics is due with the publisher. You might think you should be able to do that for real numbers too, but you cannot, so stop.

Real Numbers

This is an even broader class of numbers than the Rationals, containing all of the numbers we use on a daily basis. This includes the integers, the Rationals, as well as numbers referred to as 'irrational' – numbers that cannot be represented by a fraction, like *pi*, or the square root of 3, or any decimal that keeps on going without terminating or falling into a repetitious pattern. This is the number world where we do most of the maths we ever do in life, but it is itself only a part of a larger world, the Complex numbers, which includes all of the Real numbers, AND all of the numbers that employ the number *i*, which is defined as the square root of -1, and which are often given the grossly unfair appellation of 'imaginary' numbers.

Riemannian Space

In Euclidean space, the object that we use to determine the magnitude of any given element (or 'norm') doesn't change from location to location. So, if you want to figure out the magnitude of (1,2) or (147, 893), you're going to employ the same formula. That is not true for Riemannian Space, which is a type of non-Euclidean space, where the norm formula changes as you move from point to point in the space, sometimes to reflect some non-uniform properties of the space itself (it was this aspect of Riemannian Space that made it such a useful tool for the creation of relativity theory a half century after its 1867 introduction).

Subgroup

Before you start in here, go up to the listing on 'groups' above and read that.

So, a subgroup is just a subset of a group that is itself a group, and inherits the operation * from the group it is a subset of. Because it is a group, it has to have all the usual group stuff in it – an identity element, and inverse elements for each element in the set. The existence of subgroups then brings up questions of whether intersections and unions of subgroups are themselves subgroups, and how to find the smallest subgroup that contains a given subset of a space.

Topology

Topology is a branch of mathematics with beginnings stretching back to the mid-eighteenth century, and which came into its vigorous own in the late nineteenth. Part of what it does is study how surfaces and other geometric objects respond when placed under various stresses. What properties of an object don't change when we bend it or crumple it, and which do? This usually involves questions of how we define the closeness of points on the surface. In Euclidean space, we usually do this with some sort of norm that allows us to calculate the distance between points, but in topology we use a more general notion, based in open sets. A topology is just a space with a given rule for how to define open sets in that space. There are criteria that rule has to meet, but once you have a topology for a space, you can start talking about which points are 'close' to a point by setting up open set neighbourhoods around it and seeing what gets trapped within those borders. You can then answer questions about what happens when you start deforming that space by investigating what happens to the open sets as they are manipulated by the deformation. And that is how you can change a goat into a soccer ball.

Vector Space

Vector spaces consist of a set of elements, combined with a field of numbers, an operation that lets you add elements in the set to make other elements in the set, and an operation that lets you scale elements in the set by multiplying them by a number from the vector space's field. There are rules that the addition element and the scalar multiplication must follow, which retain the flavour of how we work with vector quantities in normal Euclidean space, but in a more generalised form, which allow us to investigate the properties of other types of related spaces by harnessing our intuitions of how things work in Euclidean space.

Building Your Own
Women in Mathematics Library:
A Selected Bibliography

General

Ambrose, Susan A., Kristin L. Dunkle, Barbara B. Lazarus, Indira Nair and Deborah A. Harkus. *Journeys of Women in Science and Engineering: No Universal Constants*. (Temple University Press, 1997).

Bradbrook, M.C. *That Infidel Place: A Short History of Girton College, 1869–1969*. (Chatto and Windus, 1969).

Brittain, Vera. *The Women at Oxford*. (Macmillan, 1960).

Byers, Nina and Gary Williams. *Out of the Shadows: Contributions of Twentieth Century Women to Physics*. (Cambridge University Press, 2006).

Cooney, Miriam P., ed. *Celebrating Women in Mathematics and Science*. (National Council of Teachers of Mathematics, 1996).

Darby, Nell. *A History of Women's Lives in Oxford*. (Pen & Sword, 2019).

Gleick, James. *Chaos: Making a New Science*. (Penguin, 1987).

Grinstein, Louise S. and Paul J. Campbell, ed. *Women of Mathematics: A Biobibliographic Sourcebook*. (Greenwood Press, 1987).

Haines, Catharine M.C. *International Women in Science: A Biographical Dictionary to 1950*. (ABC-CLIO, 2001).

Henrion, Claudia. *Women in Mathematics: The Addition of Difference*. (Indiana University Press, 1997).

Ogilvie, Marilyn Bailey. *Women in Science: Antiquity through the Nineteenth Century. A Biographical Dictionary with Annotated Bibliography*. (Massachusetts Institute of Technology, 1986).

Ogilvie, Marilyn and Joy Harvey, ed. *The Biographical Dictionary of Women in Science: Pioneering Lives from Ancient Times to the Mid-20th Century*. 2 vols. (Routledge, 2000).

Osen, Lynn M. *Women in Mathematics*. (Massachusetts Institute of Technology, 1974).

Perl, Teri. *Math Equals. Biographies of Women Mathematicians + Related Activities.* (Addison-Wesley, 1978).

Stephen, Barbara. *Girton College, 1869–1932.* (Cambridge University, 1933).

Strohmeier, Renate. *Lexikon der Naturwissenschaftlerinnen und naturkundigen Frauen Europas.* (Verlag Harri Deutsch, 1998).

Biographies, Memoirs, and Monographs

Arianrhod, Robyn. *Seduced by Logic: Emilie Du Chatelet, Mary Somerville and the Newtonian Revolution.* (Oxford, 2012).

Aschbacher, Michael, Don Blasius and Dinakar Ramakrishnan, ed. Olga Taussky-Todd: *In Memoriam.* (International Press, 1988).

Birman, Joan S. *Braids, Links, and Mapping Class Groups.* (Princeton University Press, 1974).

Cartwright, Mary Lucy. *Integral Functions.* (Cambridge University Press, 1962).

Chung, Fan R.K. *Spectral Graph Theory.* CBMS Regional Conference Series in Mathematics Number 92. (American Mathematical Society, 1997).

Deakin, Michael A.B. *Hypatia of Alexandria: Mathematician and Martyr.* (Prometheus Books, 2007).

Ehrenfest, Tatiana and Paul Ehrenfest. *The Conceptual Foundations of the Statistical Approach in Mechanics.* Transl. Michael J. Moravcsik. (Cornell, 1959).

Johnson, Katherine. *My Remarkable Journey: A Memoir.* (HarperCollins, 2021).

Mazzotti, Massimo. *The World of Maria Gaetana Agnesi, Mathematician of God.* (Johns Hopkins, 2007).

McDuff, Dusa and Dietmar Salamon. *Introduction to Symplectic Topology.* (Oxford, 1998).

Musielak, Dora E. *Prime Mystery: The Life and Mathematics of Sophie Germain.* (University of Texas, 2015).

Neumann, Hanna. *Varieties of Groups.* (Springer-Verlag, 1967).

Polubarinova-Kochina, Pelageya. *Theory of Ground Water Movement.* Transl. J.M. Roger de Wiest. (Princeton University, 1962).

Neunschwander, Dwight E. *Emmy Noether's Wonderful Theorem.* (Johns Hopkins University Press, 2011).

Reid, Constance. *Julia: A Life in Mathematics.* (Mathematical Association of America, 1996).

Riley, Rachel. *At Sixes and Sevens: How to Understand Numbers and Make Maths Easy.* (HarperCollins, 2021).

Rudin, Mary Ellen. *Lectures on Set Theoretic Topology.* CBMS Regional Conference Series in Mathematics, Number 23. (American Mathematical Society, 1975).

Scott, Charlotte Angas. *An Introductory Account of Certain Modern Ideas and Methods in Plane Analytical Geometry.* (Macmillan and Co, 1894).

Senechal, Marjorie. *I Died for Beauty: Dorothy Wrinch and the Cultures of Science.* (Oxford University Press, 2012).

Uffink, Jos, Giovanni Valente, Charlotte Werndl and Lena Zuchowski, ed. *The Legacy of Tatjana Afanassjewa: Philosophical Insights from the Work of An Original Physicist and Mathematician.* (Springer, 2021).

Vray, Nicole. *Catherine de Parthenay, duchesse de Rohan: Protestante insoumise,* 1554–1631. (Perrin, 1998).

Wieland, Christoph Martin. *Theano: Briefe einer antiken Philosophin.* (Reclam, 2010).

Dale's Favourite Books for Getting Started with Different Areas of Maths

Abstract Algebra: *Abstract Algebra* by David Dummit and Richard Foote. John Wiley & Sons.

Calculus. *Calculus: Early Transcendentals* by James B. Stewart. Cengage Publishing. (There is *no* reason to get the most recent edition that Cengage is trying to sell for the outrageous, should-be-illegal price of $300 a unit. Just grab an old used copy of the 2nd or 3rd edition and you'll be good!)

Classical Analysis: *Elementary Classical Analysis* by Jerrold Marsden & Michael Hoffman. W.H. Freeman & Company.

Differential Equations: *A First Course in Differential Equations with Modeling Applications* by Dennis G. Zill. Brooks/Cole.

Linear Algebra: *Linear Algebra and its Applications* by David C. Lay. Addison-Wesley.

Number Theory: *Number Theory, Step by Step* by Kuldeep Singh. Oxford University Press.

Real Analysis: *Real Mathematical Analysis* by Charles Chapman Pugh. Springer-Verlag.

Topology: *Topology* by James R. Munkres. Prentice Hall.